MW00511267

OUT OF PLACE

SOCIAL IDENTITIES

General Editors: Shirley Ardener, Tamara Dragadze and
Jonathan Webber

Based on a prominent Oxford University seminar founded over two decades ago by
the social anthropologist Edwin Ardener, this series focuses on the ethnic, historical,
religious and other elements of culture that give rise to a social sense of belonging,
enabling individuals and groups to find meaning both in their own social identities
and in what differentiates them from others. Each volume is based on one specific
theme that brings together contemporary material from a variety of cultures.

OUT OF PLACE

MADNESS IN THE HIGHLANDS OF PAPUA NEW GUINEA

Michael Goddard

Berghahn Books
New York • Oxford

First published in 2011 by
Berghahn Books
www.berghahnbooks.com

Library of Congress Cataloging-in-Publication Data

Goddard, Michael (Michael Bruce)
 Out of place : madness in the highlands of Papua New Guinea / Michael
Goddard.
  p. cm. - (Social identities ; v.6)
 Includes bibliographical references and index.
 ISBN 978-0-85745-094-4 (hardback : alk. paper) - ISBN 978-0-85745-095-1
(ebook)
 1. Papuans-Papua New Guinea-Western Highlands Province-Psychology.
 2. Papuans-Papua New Guinea-Western Highlands Province-Mental health.
 3. Papuans-Papua New Guinea-Western Highlands Province-Social conditions.
 4. Psychiatry, Transcultural-Papua New Guinea-Western Highlands Province.
 5. Ethnopsychology-Papua New Guinea-Western Highlands Province. 6. Western
Highlands Province (Papua New Guinea)-Social conditions. I. Title.
 DU740.42.G615 2011
 305.9'0840899912-dc22

                                                              2011000703

British Library Cataloguing in Publication Data

A catalogue record for this book is available from the British Library

Printed in the United States on acid-free paper.

ISBN 978-0-85745-094-4 (hardback)

To the Kakoli

CONTENTS

PREFACE

I have no psychiatric qualifications, and no training in that profession. Neither have I, nor anyone else in my family, ever been diagnosed as mentally ill. My research on madness in a community in the Papua New Guinea highlands was not prefigured by any longstanding preoccupation with psychiatry. Rather, it eventuated from my discontents as an anthropology undergraduate at a time when the societies we studied were still being presented in lectures as if their members were behaviourally homogeneous. My occasional queries about the analytic value of considering individuals who might be 'different' were hardly embraced, by most of my lecturers, as constructive tutorial contributions. One told me rather stiffly that anthropology should not be concerned with 'individuals'. Others referred me beyond the sacred ground of anthropology to the sociology of deviance, via a route legitimated by a modicum of shared Durkheimian ancestry. I remained unsatisfied by these types of responses as I graduated and embarked on a master's degree.

A job during a university holiday period proved to be academically portentous. Through the lottery of available short-term work on a campus notice board I found myself employed by a community mental health centre to obtain feedback from clients about the service it provided. Most of the clients were people who had been diagnosed as mentally ill, had spent time in psychiatric hospitals, and were now self-medicated and coping in the community with support from the centre. In the few weeks of my employment I obtained a basic knowledge of the procedural and pharmaceutical aspects of psychiatry, and learned something of the lifeworld of people who had been diagnosed with conditions such as schizophrenia and manic depression, as well as lesser mental disorders. Staff at the centre as well as clients – for many of whom the acquisition of psychiatric knowledge was clearly part of their attempt at coping with their condition – contributed to my expanding interest in the subject.

Pursuing an understanding of my work experiences I read a wide range of literature on total institutions such as prisons and hospitals, including psychiatric hospitals, and critiques of the attribution and treatment of mental illness, all of which aided my appreciation of the responses of the mental health centre's clients whom I interviewed. When the time came to choose a topic for a master's thesis, with my undergraduate querulousness undiminished, a potential research field was of course ready-to-hand. The community centre clients with whom I had become acquainted were willing to talk at length and the staff encouraged my project: a social study of people who had been diagnosed with psychosis, individuals who were 'different'. Among other things I collected life histories of the relatively self-sufficient ex-psychiatric patients, a project to which they said their contribution was cathartic in itself; surprisingly to me, their life histories had never been explored in depth during their engagement with institutional psychiatry. I asked about their childhood, the events which had drawn them to the attention of psychiatrists, their treatment and its aftermath. It was a sobering and, overall, depressing topic but the thesis (restricted-access and never published) was well received by my examiners.

My subsequent PhD research was intended to be a natural progression from the MA research: a cross-cultural extension of my study of the social experience of 'mentally ill' people. However in the highlands of Papua New Guinea, doing participant observation fieldwork among a people who knew nothing of psychiatric discourse or the concept of 'mental illness' at its core, I very quickly had that disjunctive experience which some of my anthropological elders had mythologized as both a revelation and a purification of the immersed fieldworker. Nothing I had learned of psychiatry and mental illness, I realized, would aid my interpretive understanding of the place of madness in this community's lifeworld (except of course as a matter of radical comparison). A corollary of this realization was that the theory and analysis which had served so well in my earlier research now appeared embarrassingly 'ethnocentric' and would be of little service. I hope this book conveys something to undergraduate or lay readers of the effects that fieldwork in a very different society can have on a Western anthropologist's comfortable understanding of his or her world.

Only a few sections of my PhD thesis have been published – as journal articles – since, following its acceptance, I immediately imposed upon it a 'restricted access' condition due to the nature of some of its content. The opportunity to revisit the substance of a dissertation for the purposes of a monograph two decades later is a luxury. One can discreetly modify what now appear to be its excesses, repair what now appear to be delinquent ellipses, reconsider the import of fieldnotes – all with the benefit of maturity, years of comparative reading and reflection, and the contemplation of changing fashions in ethnography and analysis. I went to the field to do 'pure' research towards the end of an era when neophyte anthropologists were enjoined to be as unobtrusive as possible in fieldwork, to attempt to do no harm, to participate as much as was locally appropriate with as little effect on the status quo as possible. My occasional minor interventions in local sociality – propelled by therapeutic impulses which were personal and not professional – were an embarrassment to me immediately afterwards; they felt like lapses in participant observation, impediments to the hermeneutic endeavour. I am less flagellant now in retrospect. One or two of those interventions are included descriptively in this book as a necessary part of my narrative of events: They also invite reflection on the demands of what used to be a fundamental research strategy of pure anthropology.

Currently, in Australia where I live, there is a burgeoning of 'applied' anthropology – particularly in relation to development aid and intervention in perceived social ills and disadvantage – which has moved me to reflect on the terms of reference, as it were, of my own research on madness in a Melanesian society. I had no mission to contribute to the 'development' of the Kakoli people, to educate them out of contemporarily inappropriate or bad 'customs', or to ensure their appreciation and adoption of the values and desires of my own society. Certainly I had no mission to assess and contribute to their 'mental health' or to aid the transcultural psychiatric endeavour. I regard myself as having been privileged in this respect: Surely, had my research been framed by one of the above imperatives, I would have achieved less than whatever insight into the Kakoli lifeworld I managed at the time. By the same token, I do not assume the authority to give advice to either the Kakoli or the psychiatric profession on how to proceed in the current climate of aid and 'capacity building'.

On reading a draft of this book, a colleague commented on what he thought to be a general negativity accompanying my references to psychiatric intervention or therapeutic effectiveness, and asked whether my implication was that it would be better to have no psychiatric intervention at all. In response I should say that I have no firm position on psychiatry in general, either 'pro' or 'anti'. I have encountered people in my own society who have clearly been helped by psychiatric intervention, and others for whom it appears that psychiatric attention has either been ineffectual or has compounded whatever problems they already had. The same generalization could probably be applied in the case of Papua New Guinea, if we accept the premise that florid madness is 'mental illness'. But that brings us back to the *problematique* at the heart of this book: Among people whose lifeworld does not include 'mental illness', and who do not have an 'ethnopsychiatry', what kind of intervention does psychiatry (which presupposes mental illness) represent, if the 'patient' and his or her community do not recognize its object? On the matter of therapeutic intervention into individual madness, then, I claim no qualification or paradigmatic ground from which to advise either the Kakoli or psychiatric professionals.

ACKNOWLEDGEMENTS

Content from some previous articles and a book chapter is reproduced with modification in this book, with permission of the original publishers: Chapter 1 substantially reproduces my article 'Bedlam in Paradise: A critical history of psychiatry in Papua New Guinea', published in 1992 in *The Journal of Pacific History*. Chapter 2 substantially reproduces my 1994 article 'A suitable case for treatment? The theory and practice of transcultural psychiatry in Papua New Guinea', in *Canberra Anthropology*. Chapter 4 includes a section modified from 'The Birdman of Kiripia: Posthumous Revenge in a Papua New Guinea Highland Community', a chapter in *Man and a Half: Essays in Pacific Anthropology and Ethnobiology in Honour of Ralph Bulmer*, edited by Andrew Pawley (1991) and published by JPS. Chapter 6 is modified from 'What Makes Hari Run? The Social Construction of Madness in a highland Papua New Guinea Society', published in 1998 in *Critique of Anthropology*.

Funding for my fieldwork in the upper Kaugel Valley, Western Highlands Province, Papua New Guinea in 1985/86 was provided by the Wenner-Gren Foundation for Anthropological Research and the New Zealand University Grants Committee. I thank Andrew Strathern, who was then head of the Institute of Papua New Guinea Studies, for affiliating me to that institution for research visa purposes. The research was conducted for my PhD dissertation, in which respect the late Nancy Bowers (*Nansi-o, kondo awili tekem*) and Max Rimoldi, were invaluable. Conversations with the late Ralph Bulmer were also productive.

A fieldworking anthropologist attempting to do justice to the principles of participant observation is a burden on any community. I thank, collectively, the Kepaka Alipe subclan in whose territory I was based and I particularly thank the following people variously for their patience, companionship, assistance and exegesis: Piniana Pundia, Andrew Pundia, *alaye* Manenge, Temal Topo, Simbili, Kerarus Tendi, Los Nema, Anna Los, Bebi Waka, Philip Noria, Tame Ralye, Silsu Karo, Matthew Tepu, Jack Sai, Su Topere, Kepa Pege, Vincent Auali and Paius Niningi.

Wilfred Moi and B. G. Burton-Bradley (both now deceased) graciously submitted to lengthy interviews in Port Moresby in 1985 concerning the history and practice of psychiatry in PNG. Robin Trompf allowed me access to all areas of the Laloki psychiatric institution, and Etai Tiamu reminisced on his long service there as a psychiatric nurse. Other people who were helpful in a variety of ways in Papua New Guinea were Michael Harari, Paul Barker, Gary Trompf, Fr Gordon Quinn, Br Andrew, Sr Judith De Montfort, Robert Head, June Head, Ruth Blowers, the late Carol Jenkins, the late Travis Jenkins, Alison Mott, Alec Buchanan, John Richens, Veronica Spooner, Joseph Tipeka, Bob Besari, and Mike Hughes.

Jadran Mimica encouraged me to write this book and provided rigorous comments on an early draft, as did James Leach. Judy Davis drew the map. Finally I thank my partner Deborah Van Heekeren for her support, indulgence and intellectual forbearance.

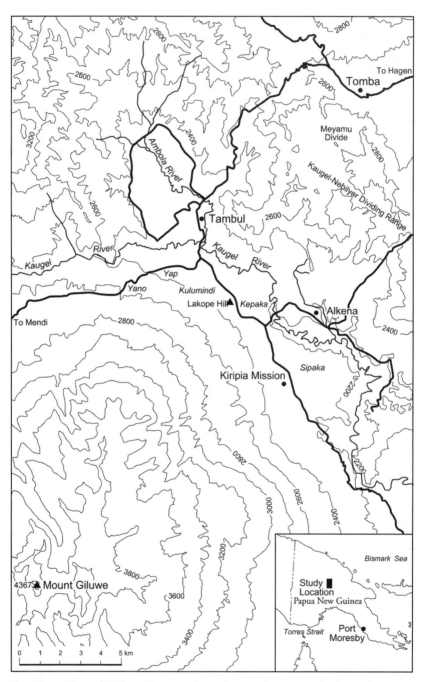

The Upper Kaugel Valley: Topography, relevant clan names (italicized), sites and features as referred to in this book.

INTRODUCTION

This book is based on anthropological fieldwork conducted in 1985/1986 in the upper Kaugel Valley, Western Highlands Province, Papua New Guinea (PNG). The fieldwork was for my doctoral dissertation on madness and psychiatry in PNG, which was never published, though I have produced a handful of related journal articles. Recently I was reminded by a colleague of the paucity of Melanesian ethnography dealing with the subject of 'mental illness' and was persuaded to review my thesis and modify it for the purposes of a monograph, more than two decades on. At the time of my fieldwork it was not yet considered inappropriate to speak of an 'ethnographic present' even if a decade or more had passed since the writer had left the field. But nowadays, in the post-everything age, it would be even more misguided than it might have been in those days to use the present tense ethnographically. The substance of this book is historical, and the reader should be under no illusion that it describes current sociality.

While the ethnographic content is more than two decades old, I believe the book has comparative and critical value. In an age when 'globalization' is part of the jargon of everyday Western commercial discourse, and Western paradigms of science and technology are increasingly being imposed on non-Western peoples in the name of 'development' and 'aid', we need to be reminded of the diversity among societies which modern anthropology once prided itself on emphasizing. A central theme of this book is that the Kakoli – the people of the upper Kaugel Valley – did not, when I conducted fieldwork, have a concept of 'mental illness'. The comparative import of this should be evident. The Western world had by the twentieth century developed a branch of medical science that addressed madness on the understanding that it was a mental disorder, and a psychiatric interpretation of madness is now more-or-less uncritically shared by the medical profession and laypeople. To acknowledge that a Melanesian people do not see madness in terms of the 'mental' means we cannot even ascribe to them a comparative anthropological category such as 'ethno*psychiatry*', since the second part of that compound presupposes a subscription to the notion of a deranged mind located in the cranial mass of nerve tissues which Westerners refer to as a brain.

A critique of the liberal intent of anthropological references to the idea of ethnopsychiatry is embedded here. A handy example of the latter is provided by Atwood Gaines (1992) who makes the laudable point that 'there is no universal psychiatric reality, no firm external base beyond culture on which stands a given ethnopsychiatry or upon which it reflects' (ibid., 5). Psychiatric systems, says Gaines, 'like religions, kinship systems, or political systems, are culturally constructed' (ibid., 3). Yet David Schneider (1984) famously said of kinship that we cannot assume that other societies have such a concept (in advance of the question of their own cultural construction of it). We should be cautious of assuming the same of 'psychiatry', given that Gaines sees the field of ethnopsychiatry to be '*mental derangements* as locally understood, treated, managed and classified' (1992: 4, emphasis added). Thus even Gaines's admirable attempt at a 'new ethnopsychiatry' going beyond the traditional focus on 'folk psychiatry' and taking as its subject 'all forms of ethnopsychiatric theory and practice whether folk or professional' (ibid., 4–5) would phenomenologically misrepresent a people such as the Kakoli, whose conception of the person did not share the physical and mental categories psychiatry takes for granted.

Madness and Mental Illness

To say that the Kakoli did not practice ethnopsychiatry or have a concept of mental illness is not to say that they did not have a concept of madness. They classified madness in social terms, as behaviour estranged from the range of responses appropriate to the social context – hence my titular phrase 'out of place' – without any necessary reference to the 'mental' condition of the person. I will elaborate on the Kakoli concept of madness and the person in chapter 3. A historical critique of psychiatry itself is implicit in the comparative observation about conceptions of madness. The origin of the concept of mental illness in the development of psychiatry has been operationally forgotten by the profession. That is to say, the idea of 'mental illness' has been disseminated and worked with for long enough that it is unreflectively treated as a universal. It is, however, a concept that developed historically within a scientistic movement in the West and became institutionalized in the nineteenth century. It subsequently came to displace other understandings of madness through more than a century of authoritative administration. Before the advent of 'mental illness', madness was attributed to a variety of social, physical and spiritual afflictions (see Doerner 1981; Foucault 1982, 2006; MacDonald 1979; Neugebauer 1981; Scull 1981).

The theoretical concerns of 'transcultural' psychiatry reflect psychiatry's professional amnesia. The problem transcultural psychiatry set for itself in the late twentieth century was to overcome the distance between Western and non-Western concepts of 'mental illness', so cultural difference and the comprehension of differing views of mental illness became the project, obviating the reflection that Western society had itself only recently developed the idea of 'mental illness'. This is a theme to which I will return occasionally during the first three chapters.

The absence of mental illness from the Kakoli lifeworld and their characterization of madness in terms of social estrangement raises further categorical problems, for the range of behaviour which they called mad was wide, from the simple contrariness of children to actions which a Western psychiatrist would identify as evidence of major psychosis. Here again, we are obliged to reconsider our categorizations of pathology. In fact, the common Anglophone term 'crazy' is far more compatible with Kakoli representations of madness than is 'mentally ill,' by virtue of its categorical imprecision in lay usage. 'Crazy' can be 'funny', 'outrageous', 'silly', and many other things – including seriously and dangerously mad.

The imprecise categorization of madness among the Kakoli presented me, when I arrived in the Kaugel Valley, with an immediate critique of what my own assumptions had been when I decided to make madness the focus of my fieldwork. While I was fully prepared for 'alternative' views of mental illness, I had not taken into account my predisposition to categorize madness as serious pathology in need of therapeutic management, conditioned as I was by my knowledge of the politics of psychopathology in the West. Madness was not always a 'problem' for the Kakoli however, and they had no systematic therapeutic strategies for dealing with it. Tellingly, my preoccupation with madness was a puzzle to them. This was partly because of their expectations about anthropological research, for after their lengthy experience of my ethnographic predecessor Nancy Bowers they thought I would be studying plants and agriculture, as she had. But their puzzlement was also due to their own understanding of, and attitude toward, madness. There was nothing to 'study' or learn about madness or madpeople, they told me. Nevertheless they indulged my preoccupation, and tried to be informative about a subject which they considered inconsequential compared to agriculture and pig management.

Ethnographically I attempted from the start of fieldwork to avoid imposing my own judgment about who was or was not mad (or 'mentally ill', for I could not at that stage deny my conditioning by Western conventional assumptions about the nature of madness). Consequently my working material was largely determined by Kakoli classifications. To begin, I collected a 'list' of people whom the Kakoli considered to be, or to have been at some time, mad. My own observations of the actions of these people, their interactions with others (including myself), the stories, rumours, gossip about them, and the spoken and exhibited attitudes of others towards them constituted much of the raw material of my research. The initial list mostly excluded minor madness, because informants naturally talked of more interesting craziness, developing narratives which were often entertaining and occasionally transparently exaggerated in the interests of providing a good story (the Kakoli considered good story telling to be a virtue). I quickly found that studying 'madness' among the Kakoli was not going to be an exercise in ethnopsychiatry. Nor would it be a collation of episodes of psychopathology explained according to a local 'culture', if culture was to be conceived in the sense still subscribed to at the time by many of my anthropological elders – a bounded entity with a stable stock of values and practices and a shared system of meanings.

Madness and Praxis

Reviewing my experiences of, and observations about, the episodes which Kakoli represented as serious, or at least spectacular, madness I was struck by the degree to which they were socially (including discursively) constructed by the madperson and the community together. I found each case to be singular, untypical and uncategorizable by either psychiatric terms of reference or a systematic notion of culture. In preference to trying to understand madness in the Kakoli community according to either of these referential contexts, then, I found it useful to examine the praxis manifest in particular episodes. As 'praxis' has come to mean many things to many people, I must specify that I use the term in a sense grounded particularly in Marx's critique of Hegel (Marx 1974: 124–47), where the dialectic of negativity – the negation of the negation – was taken to be the 'moving and generating principle' (ibid., 131) determining that no element of social reality could be investigated as complete in itself, but should be seen as an element of a social totality involved in a process of historical change.

I need to elaborate further on this, for 'dialectic' itself has been subject to many representations. The term has been applied to the Socratic dialogues, the ideal system of Hegel, the critical reformulations of Hegelian philosophy by Feuerbach and Marx and a number of subsequent models (see, for example, Colletti 1973: 7–27, 40–51; Larrain 1986: 32–8; Sartre 1976: 1–41; Wartofsky 1977: 7–18). The common description of dialectic in abstract terms, particularly in respect of its Hegelian and post-Hegelian form, represents it as a system of dynamic contradiction. A popular way of explaining this principle uses a triadic formulation whereby a 'thesis' is contradicted by an 'antithesis' resulting in a 'synthesis' which is itself a new 'thesis' and is consequently contradicted by a new 'antithesis' resulting in another 'synthesis', and so on. This is a fairly primitive model, though it has some explanatory use as an introduction to Hegel's method and in relation, for example, to reading a Socratic dialogue such as the *Theaetatus* where the process is clearly visible. However, if applied when considering humans in a historical context it suffers from an implication of mechanistic determinism. This deterministic quality is sometimes attributed to a form of Marxism known as dialectical materialism or 'diamat' and decried as crude or unsophisticated by many Marxist scholars (see, for example, Larrain 1986: 32–8; Sartre 1976: 15–22).

The diamat version is not entirely countered by formulations such as Alfred Schmidt's which hold that the dialectical process conceived historically involves 'the mutual interpenetration of nature and society as it takes place within nature, conceived in its widest sense as the total reality comprising both moments' (1971: 16). In Schmidt's interpretation of Marx, nature *becomes* dialectical by producing people as 'transforming, consciously acting subjects confronting nature itself as forces of nature' (ibid.), and as a consequence, 'Since this relationship between man and nature is the precondition for the relationship between man and man, the dialectic of the labour-process as a natural process broadens out to become the dialectic of human history in general' (ibid.). This kind of formulation, while attempting to rescue Marx from diamat appropriation, leaves some ambiguity of interpretation and does not seem to me to entirely represent the spirit of Marx's own assertion that '[t]he first premise of all human history is, of course, the existence of living human individuals. Thus the first fact to be established is the physical organization of these individuals and their consequent relation to the rest of nature' (Marx and Engels 1976: 37). Here and elsewhere in his articulation of the difference between his own position and those of Hegel and Feuerbach, Marx makes it clear that his dialectical starting point is the conscious and reflective activity of humans (see, for example, Marx 1974: 136–46; Marx and Engels 1976: 41–51).

In this respect I concur with Jean-Paul Sartre's corrective formulation, acknowledging the vital part played by conscious human activity:

> [T]he dialectical movement is not some powerful unitary force revealing itself behind History like the will of God. It is first and foremost a *resultant;* it is not the dialectic which forces historical men to live their history in terrible contradictions; it is men, as they are, dominated by scarcity and necessity, and confronting one another in circumstances which History or economics can inventory, but which only dialectical reason can explain. (Sartre 1976: 37)

Sartre attempts here to remain faithful to Marx's placement of human mental and physical labour at the core of history whereby humans develop their material production and interaction, altering 'along with this their actual world, also their thinking and the products of their thinking' (Marx and Engels 1976: 42). History, in these terms, is not dialectically predetermined, and the possibilities of its development remain flexible. As Larrain succinctly puts it, 'In realizing themselves as practical beings, men do not follow a necessary plan. So the sense of history must be given in the practical process whereby human beings realize themselves' (1986: 119–20).

In the context of the dialectical movement of human history, 'praxis' becomes more than simply a substitute term for 'practice' or 'practical activity' in the Aristotelian sense. It is problematized as the mutually transformative process in which humans, who exist both objectively and subjectively, engage with their immediate world. By 'praxis', then, I mean the dialectic (the contradictory engagement of subject and object) in its human, practical modality, the movement of the (human) subject from objective to objective which is simultaneously a surpassing and an internalization.[1] By considering the praxis involved in episodes of madness, I attempt here to understand those episodes not as cases of individual psychopathology but as part of the dialectics of Kakoli sociality, insofar as the Kakoli are a dynamic, imaginative and reflective community, actively and discursively creating and recreating themselves through time: They are, after all, a society 'with history'.

People with History

Anthropologists have a duty to endeavour to overcome the still-popular view of peoples such as the Kakoli as having been 'people without history' (Wolf 1982) in pre-colonial times. That is to say, we must avoid and discourage portrayals of them as having been static, discrete, isolated and unchanging groups for several millennia before the arrival of Westerners who brought 'social change' and thereby the beginnings of a 'history' which they did not previously have. It is true enough that the 'pre-contact' Kakoli themselves did not share the particular linear, chronological view of the past which typifies conventional Western historical consciousness. The latter is an imprecise concept: In its broadest sense it could perhaps be used to refer to a community's consciousness of its history, though this can be theorized from a number of perspectives (see, for example, Seixas 2004), and begs the question of what constitutes 'history'. A conventional anthropological view of the development of the concept of history is that it followed the emergence of literacy, which enabled systematic listing, cataloguing, and so on (Goody 1987; Goody and Watt 1963; Lévi-Strauss 1968: 258–62). Implicitly here, the consequent ability to abstractly measure time (in contrast to a previous ability only to recognize temporal sequentiality) is integral to historical investigation of the past. But 'history' itself, as Lévi-Strauss pointedly remarked in his extended commentary on Sartre, is 'never history, but history-for' (Lévi-Strauss 1968: 257), which in the anthropological context assigns 'the metaphysical function of Other to the Papuans' (ibid., 258n).

Thus a paradox is found beneath Eric Wolf's well-meaning trope. Peoples such as the pre-European-contact Kakoli did not think about their spatio-temporal conditions in a way that we would call 'historical', and in discussing their pre-contact 'history' (which becomes 'pre-history' according to documentary convention) we unavoidably draw them into a spatio-temporal interpretation that they did not share. This becomes even more problematic when we consider that the history Wolf (1982) wanted to acknowledge was of the *world* as we have come to know it, which is certainly not the 'world' the Kakoli knew.[2] Acknowledging this, and the historical consciousness from which I cannot escape by virtue of my own spatio-temporal location, I nevertheless regard as a greater injustice the characterization of the Kakoli or any Melanesians as people who spent countless millennia in small, territorially isolated groups whose lives and circumstances never changed, and whose ignorance and lack of creativity trapped them in a stone age which was immutable until the arrival of white people. This view continues to be perpetuated in the popular media, and many Melanesians have been persuaded by the cumulative effect of colonial imagery, colonial-period anthropological representation and journalistic tropes into believing that they were disturbed from a centuries-long stone-age coma only by their recent encounter with Europeans.

In respect of the pre-contact Kakoli, while we cannot ascribe a historical consciousness to them, we can certainly dispense with the view that they were an unchanging society, and that events and changes which would amount to 'history' in our reflections on our own past never occurred. The presence of incomplete and fragmentary evidence of many kinds such as myths, archaeological findings, examinable abandoned sites of agricultural activity, and the visible signs of human impact on the biosphere, are enough to alert us to an existence which is recognizable as having been 'historical' according to the criteria we apply in Western reflection on the past. The Kakoli, as my anthropological predecessor Nancy Bowers showed (1968), had not been living unchanged in the upper Kaugel Valley for millennia; they established themselves there only a few hundred years before the present. They fought and traded with newly met strangers, fled from periodic devastating frost and returned, all before contact with Europeans (see chapter 3). Their subsequent encounter with European colonizers, missionaries and capitalism needs to be understood as a continuation of several centuries of historical change. While it should not be denied that the coming of Europeans resulted in very great transformations of the Kakoli lifeworld, these did not amount to a radical 'break' with the past.

I do not subscribe to a current trend in anthropological literature to refer to everything introduced into Melanesia through contact with Westerners collectively as 'modernity'.[3] While I have no objection to the use of related terms such as 'modern' and 'modernization' in discussions of PNG, I do not believe that societies such as the Kakoli can be said to have reflectively experienced 'modernity'. The intellectual project embedded in the distinctly Western movement of modernity has hardly been visited upon PNG. For example, the varieties of Christianity disseminated in the country still represent the religion in a form relatively unmodified by the intellectual critique that was an essential part of modernity. Re-evaluations of their own traditions by the Kakoli due to their experiences of Western impositions, furthermore, were not of the same order as those that were involved in the generation of modernity in the West. Nor did they approximate the kinds of intellectual re-evaluations implicit in the autobiographical notes guiding the influential discussion of modernity 'at large' by Arjun Appadurai (1996), who grew up in Bombay, India.

I disagree therefore with formulations such as Margaret Jolly's which collectivize 'the dramatic transformations consequent on the influences of missionaries, commodity economics, and the development of states – the combination of forces that we might, after Knauft (1997), call "modernity"' (Jolly 2001: 190–1). And I also disagree with the position of Bruce Knauft, for whom 'modernity' among the Gebusi (of PNG) does not simulate that of the West but is an 'alternatively modern point of view', a phrase Knauft uses to accommodate 'experiences that pulse with ambivalent desires, contradictory goals, and conflicting images' (Knauft 2002: 19). I do not consider the confrontation with 'modern' institutions and the desire for 'modern' things to be in themselves constitutive of lived modernity. Neither do they substantiate an 'alternatively modern point of view'. The lack of a lived modernity is indicated, for example, by the enduring participatory experience of the natural environment by Melanesians. Their sense of 'belonging to the land'[4] contrasts with the 'alienation' from nature that was a constituent of modernity in the West, where the modern subject had come to experience nature as an observer (manifest, for instance, in the development of landscape painting in the nineteenth century).

I prefer, pedantically and perhaps a little clumsily, to avoid using a collective term of reference for Melanesian encounters with Christianity, colonialism and capitalism and its adjuncts (including its medical-scientistic attendant, psychiatry). In the case of the Kakoli, as with most societies in PNG, there have been discrepancies between their expectations and their experiences since their qualified capitulation to the exigencies and promise of Christianity, the administrative agencies of the State, law enforcement, the cash economy, and Western education. Yet, as we shall see in this book, their acceptance of these various impositions had not, when I conducted my research, been matched by their rejection or abandonment of corresponding constituents of a 'pre-contact' or 'traditional' lifeworld. The mismatch between the medical-scientific project of psychiatry in PNG and the ambivalent use of psychiatry by a people who do not share one of its fundamental premises is (in keeping with my dialectical theme) perhaps an indicator of a wider social complexity.

Kinship and Capitalism

Representing the Kakoli as a people 'with history', and incorporating Marx's emphasis on the centrality of human mental and physical labour in the development of material production and interaction, I employ a conventional Marxian notion that societies can be characterized in terms of their 'mode of production'. A given mode of production is primarily defined according to its particular 'forces of production' (that is, the predominant way human labour is applied to the material environment), in combination with particular 'social relations of production' (that is, the predominant social relations through which production is organized). Marx's characterization of the capitalist mode of production in the nineteenth century thus identified the industrial system as the forces of production, and the relationship between capitalists (the owners of the means of production) and labourers (who were obliged to sell their labour power to the capitalists in order to have access to the means of production) as the 'social relations of production'.

Marx's concentration on the analysis of the capitalist mode of production, and a relative paucity of detailed information on the nature and organization of social labour in non-capitalist societies of his time, resulted in an untidy representation of non-capitalist modes under a variety of titles (see, for example, Marx 1973; Marx and Engels 1976; Hobsbawm 1964). I find Wolf's (1982) revision of this variety into a model of three modes anthropologically useful. As Wolf said, 'The utility of the concept [of modes of production] does not lie in *classification* but in its capacity to underline the strategic relationships involved in the deployment of social labor by organized human pluralities' (ibid., 76). Wolf retained the general model of the capitalist mode of production more or less as originally formulated by Marx. At the same time he emphasized its important dual character: It can develop internally and expand globally, and it also has the ability to 'enter into temporary and shifting relations of symbiosis and competition with other modes' (Wolf 1982: 79). The other two modes in Wolf's model are the 'tributary mode' (incorporating what Marx had seen as the 'feudal' mode as well as the 'Asiatic' mode) and the 'kin-ordered' mode (Wolf 1982: 79–96).

Wolf's formulation of the 'kin-ordered' mode of production is useful for consideration of the Kakoli lifeworld before European contact, and the changes that occurred when the kin-ordered mode became articulated with the capitalist mode in the twentieth century. The concept of the kin-ordered mode recognises kinship as 'a way of committing social labor to the transformation of nature through appeals to filiation and marriage, and to consanguinity and affinity' (ibid., 91). In the anthropology of PNG the complexities of interaction based on kinship relations and the ubiquitous system of 'gift exchange' through which such ties are especially manifest have been a pre-occupation for almost a century since Malinowski's original description of Trobriand *kula* expeditions and his attempt to typologize relationships of obligation (Malinowski 1966: 166–94 *passim*). The expression of relations of obligation through gifting is a characteristic acknowledged in the common anthropological typification of Melanesia as traditionally having a 'gift economy'.

The gift economy has continued to prevail despite capitalism's extension into Melanesian societies. According to Gregory (1982: 116–7) its prevalence is materially based on the persistence of clan ownership of land, and hence clan-based social organization. Consequently the impact of ideologies associated with the commitment of labour to commodity production has been relatively weak. In particular the resilience of attitudes grounded in kin-ordered sociality has impeded the individualization of Melanesians – and thus their alienation from each other – by capitalist production from developing to the degree that has occurred among Westerners. In fact, writes Gregory, the introduction of capitalism has resulted not in the destruction of a traditional economy but in the development of an 'ambiguous' economy, where 'things are now gifts, now commodities, depending on the social context' (ibid., 117). He goes so far as to argue that the gift economy actually 'effloresced' (ibid., 166) in the encounter with capitalism. These observations indicate that Melanesians have appropriated commodity exchange into their own sociality, and that the interaction of the capitalist mode of production with the kin-ordered mode does not simply entail a progressive destruction and replacement of the latter by the former. While the expanding nature of capitalism might be said to make it the dominant mode, the kin-ordered mode is not a passive receptor and the articulation of the two modes is complex. As Wolf said, a useful attribute of the concept of mode of production 'is precisely that it allows us to visualize intersystemic as well as intrasystemic relationships' (1982: 76).

In societies where extended kin groups ('clans') are the major units of production, consumption and exchange, a fundamental moral imperative is generated from the need for co-operation. This imperative is visible in, for example, strong group identification and a sense of obligation to share resources. It underlies an ideation of group responsibility for the action of individual members in relation to people of other groups, and group entitlement to compensation for individual misfortune at the hands of others. While the socio-economic activities and attitudes of the Kakoli changed under the impact of colonial administration, missionization and the cash economy, communalism remained an overwhelming ethos. Their fierce descent-group loyalty and rigorous adherence to obligations to share resources and reciprocate gifts and assistance quickly impressed itself on me during my fieldwork. An analysis which contextualizes Kakoli perceptions of madness in their reflective praxis, engendered in its turn by the exigencies of kin-ordered production and exchange, must incorporate an acknowledgement of the moral imperative embedded in their ethos.

The Structure of the Book

It can be inferred from the foregoing outline of my approach to fieldwork and my theoretical framework that this book is not reductively a study of madness *per se*. Rather, in successive chapters I attempt to reveal something of the Kakoli lifeworld phenomenologically through descriptions and considerations of how they represented and reacted to what they considered to be mad behaviour. Madness among the Kakoli is, as it were, an ethnographic vehicle, rather than an anthropological subfield of the order of, say, ethnopsychiatry. Certainly this book is not eligible for classification into the anthropology of psychiatry which one writer has defined for the new century as 'an effort aimed at connecting psychiatric institutions and expert knowledge to aspects of social life, individual and collective identity, the social epidemiology and phenomenology of social suffering and so on' (Young 2008: 299). The first two chapters do, however, discuss the theory and practice of psychiatry in relation to PNG, partly to illustrate the schism between Western ideas about madness and those to which the Kakoli subscribed. These chapters also show the paradoxes which develop when an institution oriented to a psychologistic belief in the notion of mental illness attempts to incorporate a recognition of (perhaps even a concession to) extra-psychological, extra-personal factors such as 'culture'.

Chapter 1 is a critical history of psychiatry in PNG grounded in a view that broader issues of social control, rather than mental health issues, determined the development of the mental health service. For most of the colonial period doctors untrained in psychiatry dealt with 'insane natives' mostly by incarceration. Psychiatry was formally established in the country during the late 1950s. I question the conventional image of progress and change in the treatment of madpeople in PNG wrought by the growth of psychiatry, the development of new facilities and the introduction of trained mental health officers. I review the progress of the mental health service in the larger context of the country's colonial history, and argue that the changes that have occurred have been superstructural. Psychiatry in practice has perpetuated an institutional system of social control developed after the Second World War by the colonial Administration.

In chapter 2 I investigate the schism between psychiatric theory and practice in PNG. In particular I critique the implication that psychiatry can be operationally sensitive to 'cultural' difference and 'cultural' factors. In recent decades conceptions and definitions of culture have changed considerably among anthropologists, and also among 'transcultural' psychiatrists. Some contemporary anthropologists have gone so far as to argue that the culture concept is analytically inadequate and should be discarded. An implication of the increasing problematization of 'culture' among anthropologists is that attempts at culturally sensitive diagnosis by psychiatrists are unrealistic. Using clinical examples from the colonial and post-colonial eras up until the mid 1980s I show that there was a contradiction between psychiatric theory and practice in PNG, and that the social control function of psychiatry predominated over its theoretical self-characterization as a culturally sensitive mental health service.

The Kakoli are introduced properly in chapter 3, beginning with a discussion of their history and sociality. Their conception of madness is examined, and some context is supplied by way of a comparative outline of their ideas about the person and the way afflictions such as illness are understood. Here also I elaborate the fundamental proposition that madness is not regarded as mental illness, and indeed, is not placed by the Kakoli in the same category as illness, negating any attempt we might make to place the discussion of Kakoli madness in the field of ethnopsychiatry or more broadly in that of ethnomedicine. We cannot abstract 'madness' among the Kakoli as if it were something that can be investigated as a field in itself. Consequently, inasmuch as the subsequent chapters could be read as an ethnography of madness, no attempt is made to seek synchronic 'cultural' circumstances or explanations for individual madness in the conventional sense. Instead, I contextualize my understanding of madness among the Kakoli in the group's praxis, keeping in mind the dialectical process through which they have reflectively created and recreated themselves through time.

Chapters 4, 5 and 6 discuss particular episodes of, and discourses about, what Kakoli regarded as madness. They illuminate not only the range of behaviour which was considered mad, and Kakoli responses to it, but also their beliefs and attitudes, and their preoccupations as they incorporated the effects of colonialism, capitalism and Christianity into their lifeworld. Chapter 4 discusses two episodes: one in which a man's madness was attributed to an affliction by his dead mother and another in which a woman's madness was linked to an encounter with an invisible entity which a Westerner might call a 'spirit'. At first sight they involved behaviour that would undoubtedly have been diagnosed as some kind of mental illness by a psychiatrist. However by attending to the historical social context of Kakoli interpretations of these episodes I show the extent to which their discourses about madpeople were also stories of themselves, their experiential values and their engagement with the unavoidable effects of colonialism and its adjuncts.

Chapter 5 discusses an episode represented to me in advance by Kakoli as an inevitable occurrence that I would witness during my stay with them. It was the yearly madness of a local man that, they said, erupted and disappeared in regular cycles. As they predicted, the man went mad during my fieldwork period. Local reactions to the onset of madness involved what I call the ambivalent use of psychiatry. That is to say, psychiatric services were employed without a subscription to the medical-scientistic understanding of madness (as 'mental illness') which psychiatry represents. In this extended description of unfolding events and responses I draw attention to how my own presence became part of the dynamic of this man's madness, as did the actions of various authorities, and I critique the Kakoli representation of it as a temporally regular cyclic phenomenon. In its detail, the description also illustrates a point made in chapter 2 about a mismatch between the medical-scientific, therapeutic project of psychiatry and the Kakoli response to what they conceived as serious madness.

Chapter 6 is the story of a man defined as dangerously mad by the Kakoli. Like the other extended descriptions in this book it reveals as much about Kakoli sociality as it does about the person whose madness they narrated. And, as with the other episodes described herein, the dialectical relationship between the person and the community is particularly salient. The community's portrayal of this man as violently and dangerously mad was not matched (in my observation) by his actual behaviour, raising questions about why he was a social isolate. The exaggerated persona that the Kakoli constructed for me before I had even met this man was in effect mythological. In a communal exercise in moral iconography, elements of his appearance and behaviour were exaggerated and fused to create a giant who was mad. In the chapter I elaborate a theme that I have already touched on in this introduction, the fundamental moral imperative generated from the need for co-operation where extended kin groups are the major units of production, consumption and exchange. The 'mad giant' was a man whose behaviour and idiosyncrasies could be imaginatively elaborated to create a living moral example. Having become socially alien to the community, he was negative testimony to the importance of aspects of social conduct regarded as morally fundamental.

I intended to return to the Kaugel Valley to follow up my initial research, but a combination of events and influences prevented me. Taking up a position at the University of Papua New Guinea (UPNG) in 1990 seemed to offer me the proximity and opportunity for ongoing research in the valley, but an early attempt to re-enter the area was aborted because of the outbreak of local warfare which prevented travel into or out of the valley.[5] Meanwhile new research interests in PNG's capital, Port Moresby, had begun to preoccupy me. Time passed, and my urban research consumed me to the degree that my only contact with the upper Kaugel was occasional letters and chance meetings with migrants in Port Moresby. Since leaving UPNG and moving to Australia in 1995 I have returned to PNG (including areas in the highlands) several times but never, alas, to the upper Kaugel. I have monitored developments in psychiatry in PNG since my fieldwork, but the Kaugel Valley remains, in this book, very much in the ethnographic past. Anything I could say about changes which have taken place there since the mid 1980s can only be generalized from my knowledge of other parts of PNG and from reading what others working more recently in the vicinity have written. Some comments in this regard can be found in my short concluding chapter, in which I revisit and gather themes in the six preceding chapters and consider some more recent literature indicating the kinds of changes that have occurred particularly in the worldview of people in the Mt Hagen area since my fieldwork period. A brief commentary on these reported changes shapes a question about what developments would turn the Kakoli view of madness into ethnopsychiatry.

Notes

1. My interpretation of the notion of praxis is influenced by Sartre's lengthy discussion in *Critique of Dialectical Reason* (1976). In that work he attempted, among other things, to analyse the relationship between individual praxis and group praxis. I have not imposed that particular complication upon the Kakoli herein.

2. Wolf's probable response to this point can be inferred from his comment that '[p]erhaps "ethnohistory" has been so called to separate it from "real" history, the study of the sup-

posedly civilized. Yet ... The more ethnohistory we know, the more clearly "their" history and "our" history emerge as part of the same history' (1982: 19). He goes on to cite Lesser (1961: 42) to the effect that we should think of all human societies, 'prehistoric' or otherwise, as open rather than closed systems – involved in weblike, netlike, connections with other aggregates near and far (Wolf 1982: 19).

3. For critical discussion of the use of 'modernity' in reference to Melanesia see, for example, Friedman (2002); Friedman and Carrier (1996); Hirsch (2001).

4. This has been repeatedly expressed to me not only by Kakoli but also by 'villagers' in other areas of the PNG highlands and on the coast that I have visited over a number of years. It was also a common expression among students at UPNG when I taught there, particularly when issues such as 'land ownership' were discussed in lectures and tutorials.

5. My former host clan, with which I had become identified in Kakoli perceptions, lived a few miles down from the head of the valley. The area between them and two road-entry points into the valley was controlled by two clans with whom they were now at war. Regardless of my own desire to get in, I was regarded as an 'ally' of one side in the war and was not permitted through enemy territory.

1

THE DEVELOPMENT OF PSYCHIATRY
IN PAPUA NEW GUINEA

In 1957 the Australian Ministry for Territories commissioned the psychologist A. J. Sinclair to survey mental health among indigenes of what was then called the Territory of Papua and New Guinea. Sinclair found almost nothing he could recognize as mental disorder among the colony's native population (Sinclair 1957: 30) but made recommendations which led to the formal introduction of a mental health service, with the appointment in 1959 of a psychiatrist to head a mental health subdivision of the colonial Administration's health department. For a decade or so this subdivision struggled to establish a psychiatric service in the territory, competing with other administrative departments for funds and other resources, until at the beginning of the 1970s PNG had a psychiatric hospital at Port Moresby and psychiatric wards at main hospitals around the country.

Ostensibly, then, the Sinclair survey divides the prehistory from the history of psychiatry in PNG. This chapter is a critical investigation of the process outlined above, grounded in a view that broader issues of social control, rather than mental health concerns, determined the development of the mental health service. While the service's colonial head, B. G. Burton-Bradley, was professionally correct in retrospectively positioning the birth of psychiatry in PNG in a galvanized-iron shed (Burton-Bradley 1975: 107) at Bomana, a swampy area just outside Port Moresby, the events of the late 1950s and the subsequent progress of the mental health service can best be understood if reviewed in the larger context of the country's colonial history. This perspective brings into question the image of progress and change in the treatment of madpeople in PNG wrought by the growth of psychiatry, the development of new facilities and the introduction of trained mental health officers. In reality the changes were superstructural, and psychiatry in practice perpetuated an institutional system of social control developed after the Second World War by the colonial Administration.

Colonial Attitudes towards 'Mental Illness' until the 1950s

In the earlier part of the twentieth century, apart from an insanity ordinance introduced in 1912 as a paper formality, there was no official concern with the sanity of Melanesians. Australia's political relationship to the country was driven largely by rationales of possession rather than of control or 'development', and the attitudes of administrative officers to the social behaviour of their charges tended to be *laissez faire,* so long as there was no threat to Europeans or their property. In this spirit, an early magistrate wrote anecdotally of 'the habit of running amok. This cheerful habit of the Papuan is a thing that breaks the monotony of station life and of expeditions in New Guinea, and he does it without the slightest warning and without cause; and, when in that cheerful frame of mind, he will slay friends or foes alike, until he is pulled down' (Monckton 1922: 152). Monckton provides one of the earliest descriptions of this alleged[1] behaviour, in an anecdote of an eventful night during a 1906 expedition. One of his carriers woke with a scream and 'violently rushed about, foaming at the mouth, and thrashing wildly with his tomahawk' (ibid.). Other carriers restrained him and Monckton put him in leg irons and handcuffs and forcibly dosed him with 'strong purgatives'. The following morning the man was perfectly normal again, and 'the matter ended. It was not a case for punishment' (ibid., 153). Monckton cites two more witnessed cases of 'amok' behaviour (all of them occurred in 1906) in which he did not intervene even though in one incident three Papuans were stabbed and the 'lunatic' was shot repeatedly and finally bayoneted to death by Papuan police constables (ibid., 152–55).[2]

Anthropologists and missionaries recorded a number of individual and group phenomena perceived as involving abnormal mental states, some of which have become classics in genre literature, such as F. E. Williams's (1976) account of the 'Vailala madness' of 1919 and 1920, but in which the Administration interested itself only as far as observing events in case there was a possibility of countrywide social disruption (ibid., 342). Interestingly, in the case of the Vailala madness Williams, the government anthropologist, was concerned at the participants' destruction of what he regarded as 'treasures of religion and art' and commented that 'compared with the wholesale destruction of native practice and tradition, the collective nervous disorder is matter of small and transitory importance' (ibid., 332).

Following the Second World War the Administration was preoccupied with reconstruction and rehabilitation, guided ideologically by a promised 'New Deal' for Papua New Guineans (see Wolfers 1975: 112–15; cf. Fitzpatrick 1980: 91–92, 145–46) that manifested itself largely as an intensification of control. In contrast to the lack of systematic policy before the war, the Australian Government increasingly busied itself with regulations and policy decisions, though until the 1950s, when the Labor Government was replaced by a Liberal-Country Party co-

alition, available funds were concentrated on reform and repair rather than economic development (Wolfers 1975: 119–23). Administration medical records in
PNG's National Archives are uninformative about the immediate post-war years
and official data are scrappy until the late 1950s. The first hint of practical medical concern with indigenous minds or brains is given in a memoir of missionary
medical activities which notes in passing that lobectomies were performed at the
'native' hospital at Ela Beach, Port Moresby, between 1948 and 1950 (Kettle
1979: 107).

Segregated hospitals persisted in PNG until 1963. The Ela Beach hospital,
built over the water, is described by Kettle as having 'a large ward 48 feet by 21
feet with a 10 foot verandah all around and with four small rooms, one at each
corner' (ibid., 18). Rubbish and organs removed in operations were thrown into
the sea. The lobectomies were performed by a Dr Alan Maruff and according
to Kettle (ibid., 107) it was the first time such operations were performed in
the country. While I could find no mention in archival health department files
that lobectomies were being performed at that time, the dates given by Kettle
belong to an era when brain cutting was modish in European psychiatry (see, for
example, Clare 1976: 272–74), and it is not unlikely that a doctor could have
introduced the technique at Port Moresby in 1948.

The establishment of permanent administrative staff at a number of places in
the interior as well as at coastal centres, accelerated by the war and the rebuilding
programme, brought a practical interest in areas of indigenous life which had
previously been ignored. Indications of this can be found in surviving health
department files containing correspondence from field officers enquiring about
asylums or resources for dealing with 'mental' problems. On 20 September 1950,
for example, the District Medical Officer at Goroka in the Eastern Highlands
wrote to the Director of Public Health, J. T. Gunther, asking for a directive on
the treatment and 'disposal' of 'mental disease in indigenous natives'. A man had
been brought to Goroka by aircraft from Kundiawa, nearly 50 km away:

> He arrived tightly lashed to a stretcher, which was no doubt necessary in view of his
> immediate history and subsequent behaviour. He appears to be either a violent para-
> noiac or some kind of manic-depressive, but personally I have neither the ability nor
> experience to diagnose and treat these cases.
>
> There appears to be no alternative for the present but to keep him lashed down to
> a bed, under paraldehyde [a sedative]. I have no doubt that this native is certifiable,
> according to European standards, but I do not know what machinery or methods of
> disposal are available in the territory.
>
> According to the ADO and Sub-Inspector here, the matter of disposal presents con-
> siderable difficulty.[3]

Gunther's reply, a fortnight later, stated that there was a grave lack of facilities
for the care of 'mentally deranged' patients due to the 'inability of the Dept of

Works and Housing to construct an asylum at Port Moresby'. He said that shock therapy was being used 'successfully' by a medical officer at Ela Beach, and suggested that the patient be sent to Port Moresby.[4] In a letter to the Government Secretary on the same date as his memorandum to Goroka, Gunther complained that an asylum had not been started despite (unspecified) reports elsewhere alleging it had been built in 1947.[5]

The process of committal at the time involved type-written medical certificates citing the Insanity Ordinance of 1912 and headed 'Form of Medical Certificate to accompany order or request for reception into an asylum or reception-house as the case may be'. In declaring that a person was of 'unsound mind and a proper person to be taken charge of and detained' the committing (European) medical officer was required to give brief evidence of insanity under two headings: '1. Facts indicating insanity observed by myself' and '2. Other facts indicating insanity communicated to me by others'. I found some surviving certificates in the National Archives in 1985. A certificate dated 19 August 1948, stated under heading 1:

This native stated that there are times when he is not responsible for any actions of his. He has a memory of events before and following criminal acts but not during the act which is usually at the time of the full moon and for indefinite periods thereafter.

and under heading 2:

I have previous knowledge of this native claiming whilst of unsound mind to be a policeman or medical orderly on which occasions he has been threatening to villagers.[6]

Another document dated 23 July 1952, apparently referring to an examination at Ela Beach native hospital, stated under heading 1:

Patient was observed by me to swim a very long way out to sea on a rough day and had to be brought in by boat. Speech meaningless at times religious with mental excitement and sexual tendencies.

and under heading 2:

Patient owing to excited state had to be kept in cell under restraint. Talked nonsense about religion: shouts loudly and very active.[7]

By the end of 1950 an 'asylum' existed at Bomana, about 19 kilometres inland from Port Moresby. It was a wire-enclosed compound with a lean-to shelter and from the time it was established it was a source of complaint. It accommodated ten patients and had no facilities for females. On 28 December 1950 Gunther stressed a need for female facilities, adding that a woman patient had complained of rape by male orderlies. He also said that as knowledge of an existing asylum spread, a request had been received to take in 'criminal lunatics'.[8] It is possible that some 'criminal lunatics' were being held at Ela Beach hospital at the time. When I interviewed Dr Wilfred Moi, the assistant secretary for mental health in PNG, in 1985 he recalled that in 1952 the Ela Beach hospital had a 'lockup' for 'aggressive' patients, to whom the barbiturate phenobarbitone was administered (probably as a sedative).[9] In 1954 the Bomana institution's official status was defined in the Territory of Papua Annual Report as 'a mental asylum for the criminally insane' (TPAR 1955: 51). A Papua New Guinean medical assistant, Etai Tiamu, who was employed at the Bomana institution in 1954, recalled in a 1985 interview that at that time there had been between forty and fifty patients in the asylum, with separate sleeping quarters for males and females, but no segregation during the day. He said electro-convulsive therapy (ECT) had been used and medication included phenobarbitone, chlorpromazine (an anti-psychotic) and paraldehyde, the latter administered by injection.[10]

There were no specialist psychiatrists in Papua New Guinea in the mid 1950s. The Bomana asylum was run by a European 'medical assistant', A. J. Prior. The Territory of Papua was represented at the Australian World Health Organization Mental Health seminar at Sydney in 1954 by the 'officer in charge, Infant and Maternal Welfare' (TPAR 1955: 51). The only legislation covering the establishment and operation of asylums was the Insanity Ordinance of 1912, unmodified from Australian law. It reflected turn-of-the-century Western attitudes to insanity, in that it contained no references to types of therapy or medical treatment (LTP 1949: 2233–80). Details of the kinds of people kept at Bomana are unknown. The asylum's records are said to have been destroyed by rats (Burton-Bradley 1975: 107).[11] Its conditions were far from satisfactory in the eyes of Prior, the man trying to operate it. He complained to the Health Department in 1954 of overcrowding, lack of staff facilities, fly problems and dysentery, a shortage of latrines, and the impossibility of cleanliness in the kitchen.[12] In 1957 he was still asking for better sex-separation facilities, citing pregnancy and rape of women patients.[13]

The most complete description of the Bomana asylum after ten years in existence was given in a health inspector's report of 1960: The patients were held in a wire enclosure measuring 200 feet by 150 feet, the sexes segregated by a building and a fence. The building was the patients' living quarters, made of timber with a concrete floor. The female compound had two single pan latrines and two showers, the male compound contained three blocks of pan latrines and showers (of which one was unused, and another was in a security area for 'special patients'). Meals were conveyed fifty yards from a cookhouse to a ten-yard by six-yard servery. The cookhouse had an old 'found-in-the-bush' wood stove. Cooking utensils consisted of one-third of a forty-four gallon drum for rice, and one sixth of a forty-four gallon drum for frying ('ordinary cooking utensils have been considered too small to be serviceable'). The beds, floors and walls in the 'security room' were 'continually fouled owing to the disposition of patients'. The grounds were badly drained and swampy. The health inspector recommended urgent attention.[14] Despite the shortcomings he described, the asylum would remain virtually unchanged until its eventual closure in 1967.

The 1950s and the Sinclair Report

During the 1950s the Australian Administration began to concentrate on development: Its paternalism was beginning to be displaced by an awareness of PNG's inevitable future decolonization (Fitzpatrick 1980: 92–96). The United Nations (UN) was increasingly agitating for infrastructural development, though E. P. Wolfers has commented, 'It is debatable whether the UN criticism of Australian territorial policy constituted a form of pressure at the time, or simply served to make Australian policy-makers aware of the contemporary incongruity of their attitudes' (Wolfers 1975: 126). In 1957 an Australian psychologist, A. J. Sinclair, was commissioned by the Administration to undertake a mental health survey of PNG's population. Two decades later it was retrospectively suggested (Burton-Bradley 1976: 4) that the survey had been organized as a result of a visit by a UN group in the early 1950s which stimulated health authorities' interest in mental health. However, the reports of the three UN visiting missions of the period (United Nations 1951, 1953, 1956) contained no references to mental health, though the 1956 mission made strong comments on the need to improve general health services (1956: 37).

It is not clear whether the Sinclair Report was part of a general practical response to the UN prompting on health facilities. Paul Hasluck, who was Australian Minister for Territories from 1951 to 1963, later wrote as though it came almost by chance:

> About this time my personal attention was drawn by two Australian psychologists, Dr A. J. M. Sinclair and Prof. D. W. McElwain, to the problems of mental health among the indigenous people and I arranged for Sinclair to make an inquiry. His report was received at the end of 1957 and referred to me by the Administration for urgent attention. Eventually, in 1959, we established a mental health service and Dr B. G. Burton-Bradley was appointed. (Hasluck 1976: 342)

The survey may have been stimulated by the general upheaval in Australian mental health services during the 1950s, when a wave of reformism and newspaper campaigns began to change psychiatric services, which were portrayed as backward and decrepit (see, for example, Bates 1977; Dax 1961).

In a report of the UN Visiting Mission of 1959, the Sinclair survey was retrospectively connected to alleged Administration concern over the implications of recent indigenous religious and political social movements, indiscriminately

glossed by the category term 'cargo cult' and seen as indicating confused thinking under the influence of European culture (United Nations 1959: 2–3). The
mission reported that that the Administration had 'endeavoured to combat the
cargo cult by rousing the interest of the people in cash crops of coconuts, coffee,
peanuts and rice. It has also recently had a psychological study made, the results
of which are contained in a report commonly known by the name of its author,
Dr Alexander Sinclair' (1959: 3). However, no support for this 'cargo cult' connection can be found in official correspondence previous to the survey. An Administration report to the UN covering the year immediately before the Sinclair
survey claimed, in fact, that there had been no major religious or quasi-religious
movement in the territory for some years, and spoke dismissively of unspecified
small scale movements reflecting wrong thinking which had 'prevailed for only a
short time' (RTNG 1958: 17).

The Sinclair Report marked the introduction of the concept of mental health
into PNG. Until 1957 the European administration had concerned itself only
with the sequestration of madness (when it came to the attention of Europeans).
The introduction of electroshock machines, sedatives and antipsychotic drugs
into the country had no connection with any organized or systematic mental
health programme. They were never represented as other than instruments of
control. Sinclair was the first professional psychologist/psychiatrist to set foot in
the country in other than an anthropological role. His survey took almost three
months. He travelled extensively, eliciting information from 'enlightened Europeans', 'educated native persons' and 'other indigenes' and compiled a thorough
report of some sixty pages covering a spectrum from urban neurosis through
acute and chronic psychosis to mass hysteria (Sinclair 1957).

Sinclair's observations and comments are discussed in the following chapter;
his importance for the current discussion of the development of psychiatry in
PNG lies in his recommendations. He advised the establishment of a mental
health organization consisting of a Commonwealth advisory committee in Australia on the mental health of Papua New Guineans, a permanent committee in
Port Moresby on mental health and cultural development, and a mental health
service for the treatment of all inhabitants of the country, indigenous, 'mixed
race' and European (ibid., 21). He also recommended mental health institutions
to include psychiatric wards at Port Moresby hospitals, 'colony settlements' for
chronic and long-stay patients within easy reach of Port Moresby, and small treatment centres attached to regional hospitals (ibid., 49–53). He urged incidentally
that Bomana cease to be the 'mental hospital' for the Port Moresby area as soon
as possible, though he had high praise for Prior: 'the most outstanding feature of
the hospital is the sympathetic and humane handling of patients by the European
Medical Assistant' (ibid., 40).

Sinclair's recommendations reflected the ideological shift in the Administration's relationship with PNG from control to care and development. The Bo-

mana asylum, established only half a decade previously and officially catalogued as 'an asylum for the criminally insane', was now considered 'a highly unsuitable institution for the treatment of native patients, characterized by the complete absence of any facilities for rehabilitation' (ibid., 40). In 1959 the Territory of Papua Annual Report (covering the period from mid 1957 to mid 1958) made, in a chapter on public health, its first reference to mental health – 'Consideration is being given to the establishment of a Division of Mental Health' (TPAR 1959: 69) – while a government report to the UN on the administration of the Territory of New Guinea[15] stated more self-consciously, 'A survey of mental health in the territory was carried out during the year by a psychiatrist and a professor of psychology from Australia. Their report, which included a recommendation for the creation of a Division of Mental Health in the Dept of Public Health, is at present being studied' (RTNG 1959: 100).

The following year the Territory of Papua report announced the existence of a Division of Mental Health whose functions included 'the investigation, treatment and prevention of mental health problems' (TPAR 1960: 72). The Bomana asylum was descriptively transformed in the same document into 'a special administration hospital' to which cases requiring electro-convulsive therapy (as opposed to drug therapy) were transferred (1960: 75). The latter statement, implying an existing system of specialized treatment, was in fact a modified extract from the Sinclair Report of two years earlier, which had stated 'There are no facilities in the hospitals for electro-shock treatment. Few Medical Officers are familiar with the intravenous sedation techniques such as "Amytal" and "Largactil". It could be said therefore that there are no facilities for adequate treatment of acute mental illness in the Territory except at Bomana Mental Hospital near Port Moresby' (Sinclair 1957: 39). The report to the United Nations for the same period (mid 1958 to mid 1959) also announced the establishment of the mental health division and added – imaginatively – 'The treatment of the mentally ill is undertaken at all Administration general hospitals' (RTNG 1960: 107).

The 1960s and the Establishment of Mental Health Services

In 1959 Dr B. G. Burton-Bradley, an Australian psychiatrist who had previously worked in Malaysia, was appointed to head mental health services in PNG with the title 'Assistant Director (Mental Health)' and immediately embarked on a one-year diploma course in social anthropology in Sydney, reportedly on his own initiative.[16] At the same time, plans were developed for a psychiatric hospital. This seems to have been initially intended to form part of a complex originally proposed in 1954 as a tuberculosis and hansenide hospital at Laloki, a site near Bomana. A letter to the Department of the Administrator dated 17 November 1959 listed some requirements of the psychiatric institution: accommodation for 300 patients (divided into 186 males and 114 females), and provision for four types of patients – criminally insane and violent, acute cases, convalescent patients, and chronic and harmless cases. The letter was unsigned but bore the typed initials of R. F. R. Scragg (then Director of Public Health) and Burton-Bradley.[17] Subsequent correspondence in 1960 omitted reference to TB and hansenide sections[18] and correspondence from Scragg to the Department of the Administrator dated 30 September 1963 mentions that TB and leprosy hospitals were administered by the London Missionary Society, who were 'unwilling and unable' to move into the mental health field. Scragg went on to suggest that TB and leprosy were 'stigma' diseases and it would be better not to associate mental-illness care with them, but to annexe it to the general hospital system to allow the public to regard mental illness 'in the same way' as any other sickness.[19]

In 1961 the Permanent Committee on Mental Health and Cultural Development which Sinclair had recommended was set up and had its first meeting. The Territory of Papua Annual Report for 1961–62 described it as 'studying trends in culture contact both in the Territory and Overseas' (1963: 106). The committee, and its orientation, was a manifestation of the ideology guiding the establishment of a mental health service in PNG. Social Darwinism pervaded the Australian Administration until the end of the colonial period, notwithstanding the presence of a number of enlightened individuals in its service. The Department of Territories' apologia to the UN visiting mission of 1953 for its lack of a long-term development plan cited one of its obstacles as being 'the backwardness of the indigenous inhabitants' (United Nations 1953: 14), and the guidelines for Lutheran missionaries warned, 'The native has very strong sexual drives which he often is unable to control' (Wolfers 1975: 135). In the face of this general colonial disposition it is not surprising that psychiatry became another locus for the consideration of indigenous backwardness and 'cultural development'. The report to the UN General Assembly on the administration of New Guinea for 1958–59 retrospectively implicated the Sinclair survey as part of a general research project into the strain being undergone by indigenes as they coped with change. It contained generalizations about the confusion, frustration and bewilderment of 'traditional' peoples in such circumstances and stated, 'In recognition of these situations information is being collected in the fields of anthropology and mental health', citing the Sinclair survey as an early example of the trend (1960: 18–19).

Social Darwinism was handily complemented by the conventional wisdom of psychiatry of the period, combining a medical model with a view that psychosis was a reaction to a stressful environment. Thus Sinclair explained the purpose of the 1957 field survey: 'The aim of this survey has been to attempt to interpret the behaviour of the indigenes as reactions to: (a) pressures inherent in their native culture; (b) pressures which have been superimposed by Europeans' actions' (1957: 2).

This approach, joined by a clinical perspective giving consideration to 'constitutional factors' including physical constitution, biochemical function and hormone balance (ibid., 31), was a direct imposition on to PNG society of a body of psychiatric opinion which has been described as 'merely an updated and sophisticated development' (Mellett 1982: 7) of theories produced in nineteenth-century psychiatry in England in the burgeoning of industrial capitalism. The concerns of physicist theorists in that earlier period had been 'not so much the fact of economic change, but rather the gradually apparent effects of that change on social and cultural and moral life' (ibid., 65). According to Klaus Doerner, nineteenth-century psychiatry not only 'supplanted the Romantic villainous role of the inner man and passions with the social environment, the "milieu" … but also led to the criticism of the existing institutions … which were to be improved by changing the external conditions of the asylums' (1981: 87).

The Territory of Papua Report in 1962 announced that work had begun on 'the first stage of a new psychiatric centre near Port Moresby which, when complete, will cater for 300 patients' (TPAR 1962:105), and in the first weeks of 1962 some buildings were completed beside the river at Laloki, and a few patients were shifted from the Bomana asylum. The transfer of patients came to an abrupt halt however within four weeks when the new hospital buildings were flooded to a depth of four feet by a seasonal rise in the river.[20] A new site on higher ground nearby was quickly found and work began again. A proportion of the first site not reached by floods was maintained and would eventually become a rehabilitation village.

While the Laloki project was proceeding, the Insanity Ordinance of 1912 was replaced by the Mental Disorders and Treatment Ordinance of 1960. The title itself indicates the change in the Administration's conception of madness, which was legally reinforced within the ordinance:

> In any law in force in the Territory or part of the Territory or in an instrument or document, a reference to a lunatic or to lunacy or to a lunatic asylum or asylum or mental hospital or a similar expression shall be read as a reference to a mentally disordered person or to a patient or to a mental hospital, as the case requires, within the meaning of this ordinance. (Mental Disorders Ordinance 1960: 13)

Much of the ordinance was inappropriate to PNG, since it was preoccupied with the legalities of dealing with a committed person's business interests, estate, funds, etc. (ibid., 4–7) – matters which were largely inapplicable in a land whose population was overwhelmingly engaged in subsistence activity. At the same time it devoted some space to patient's rights and hospital visitors (ibid., 8–13). It also included a section covering the certification of 'A Medical Assistant, Assistant Medical Practitioner, Nurse or Nurse (Territorial)' to administer electro-convulsive treatment in cases where the treatment could not be given by a legally qualified medical practitioner (ibid., 13). This section enabled Burton-Bradley, in the absence of professional staff at the time, to organize a system under which ECT could be administered without the presence of a qualified doctor. Assistant Medical Practitioners were issued with a certificate of proficiency after undergoing training in the operation of ECT machines and passing an oral examination.[21]

In contrast to the speed of construction of the first Laloki installation – it took about a year – the new establishment on higher ground, referred to in official correspondence as Laloki II, took six years to complete. Archival remnants imply that Scragg and Burton-Bradley had difficulty persuading the Administration to provide more funds, though no specific record of this could be found.[22] The Bomana asylum was still being used during the early 1960s, and Prior was urging Scragg to close it.[23] During this period Scragg wrote to the 'Assistant Administrator (Services)', Department of the Administrator, probably in an attempt to speed up the building of Laloki II, mentioning a past reliance on villages to care for the 'mentally ill' and commenting:

> However, despite commonly expressed opinion the village in no way represents the ideal location for the patient whose conduct is overtly abnormal. He is subjected to constant torment from exuberant village children and is often goaded to acts of violence. When such violence alarms the village, he may be caged, and whilst this effectively removes him from circulation, his mental anxiety and deterioration increase.

He added the claim that detribalization and urban drift would create strains increasing the need for institutional care.[24] Efforts to have the Laloki institution completed were boosted in late 1964 when a European Catholic Sister at Kandrian, New Britain, was murdered by an 'insane native' with an axe. Kokopo town councillor H. Hoerler asked what the Administration was doing to protect people from 'maniacs like this', and a Mr Neal said at a council meeting that he had been told by an Administration doctor that 'it was extremely difficult or virtually impossible to give native people psychiatric treatment, they were just unable to respond to it'.[25] The day after Neal's remarks were reported in the *South Pacific Post* newspaper, Burton-Bradley sent a letter to the Acting Director of Public Health denying that 'natives could not be treated', and cited the report in urging the immediate building of Laloki II.[26]

The Territory of Papua report, meanwhile, was representing psychiatric services as adequate and improving, announcing that treatment at administration hospitals 'takes into account cross-cultural factors, and includes in addition, ataractic therapy, psychiatric social work and occupational therapy' (TPAR 1965: 99). There were elements of truth in the statement: An occupational therapist and a social worker had been appointed to the mental health division in 1963 and Burton-Bradley had inaugurated the training of psychiatric nurses at Bomana, where Prior had previously had to teach his assistants on the job. Yet as late as 1965 the Controller of Corrective Institutions was corresponding with the Acting Secretary for Law, expressing his unwillingness to accept people suffering from mental illness at Bomana gaol and referring to possible damage to the Administration's reputation if the press found out that mentally ill patients were being held in gaols.[27] This prompted an immediate communication from the Acting Secretary to the Administrator that 'dangerously insane' people were being held in the maximum security section of Bomana gaol.[28] Meanwhile Prior, who had managed to overcome the rape and pregnancy problems at the Bomana asylum by shifting all the female patients to the operable parts of the original Laloki complex, was citing the seasonal flooding of the river as a further factor necessitating the completion of Laloki II.[29]

Laloki II was finally completed at the beginning of 1967, and officially designated as a mental hospital by order of the Director of Public Health on 28 February 1967 (Laloki I was thereafter the 'rehabilitation annexe'); Bomana asylum was closed at the same time.[30] The official role of PNG's psychiatric services was described at the time in the Territory of Papua report: 'The Mental Health Division advises the Administration on preventive aspects of mental health and forensic ethnopsychiatry, undertakes cross-cultural psychiatric research and studies trends in acculturation and community mental health in the Territory and overseas'. Two priorities of the mental health services were: '(a) to direct social and psychiatric resources to the high incidence of recently urbanized groups; and (b) to direct attention to members of the community who may be able to play the role of "community caretakers" in community health services' (TPAR 1969: 79). No practical action was in fact taken on the second priority until 1985, ten years after PNG's independence.

The 1970s: Ideals and Realities

At the time of Laloki's completion, PNG's main hospitals had ECT machines, supplies of chlorpromazine and a room for the mad. In 1964 the Administrator had established the general hospitals at Lae, Rabaul, Madang, Wewak and Goroka as 'mental hospitals and admission centres' under the 1960 Mental Disorders and Treatment Ordinance.[31] A report to the UN in 1971 officially represented Laloki hospital as an 85-bed institution providing psychotherapy and modern psychotropic drugs, and recreational therapy (including swimming, football, basketball, quoits, cricket, table games, bus picnics and cinema trips). It was said to have a rehabilitation annexe 'designed along village lines with a farming and garden atmosphere', a workshop and patient-run halfway house, and ready access for families of patients, with an objective to increase family therapy (RPNG 1972: 169).[32]

In contrast to the idealized version given to the UN, a psychologist who worked at several centres in PNG from 1974 to 1979 wrote in the latter year that

A visit to most any psychiatric facility in P. N. G. will reveal that occupational therapy is rarely practised. Despite the past provision of sending a number of psychiatric nurses to Australia to learn the rudiments of conducting occupational sessions, it is clear that P. N. G. psychiatric nursing staff have little inherent interest in propagating these activities. (Robin 1979: 35)

The writer alleged that during his working period mental health authorities did not attempt to make psychotherapy available to patients, pointing out that the pro-Western context of psychotherapy did not fit within the 'highly contrasting culture' of PNG, and that communication and language difficulties prevented the consideration of 'talk-therapy' (ibid., 36). He added that there was only one social worker to deal with all mental patients at both Laloki psychiatric centre and the psychiatric ward at Port Moresby General Hospital. In most other locations in the country, psychiatric wards had to share the services of the social worker with the rest of the hospital, and the social worker was overworked and most frequently unable to get to the patients' home areas due to lack of time and transport (ibid., 37). Finally, in contrast to the 'official' comment that families of patients had ready access, Robin pointed out that the physical isolation of the Laloki psychiatric centre from most patients' village communities mostly prevented visits from family and friends, and further,

Many patients have been functioning within socially acceptable limits for several years but due to prolonged institutionalization have lost contact with their homes. As a result, when an effort is made to socially re-integrate these patients, many of them have a hard time fitting back into their home settlements. (ibid., 52)

Published ten years after the report to the UN which they largely contradicted, Robin's comments were part of a booklet representing the critical extreme of the reassessment of psychiatry in PNG which came with the end of direct colonialism in the mid 1970s. In 1975, when the country gained political independence, Burton-Bradley was replaced as Assistant Director, Mental Health, by Dr Wilfred Moi, a Papua New Guinean whose training in psychiatry Burton-Bradley had largely guided. Until that time Burton-Bradley had complete control over mental health services, and no study of psychiatry in the country can avoid some discussion of him and his contributions. His theoretical contributions (which still dominate PNG psychiatric literature) will be discussed in the following chapter. Here, though, we are concerned with his part in the development of psychiatry in PNG.

Throughout his sixteen years as the head of psychiatric services Burton-Bradley never altered or modified the medical model he brought to his position in 1959. His approach is clearly represented in a paper titled 'The Psychiatric Examination of the Papua and New Guinea Indigene', written in 1965, which he opens with the following statement:

The medical officer in the field, confronted for the first time with a suspected case of psychiatric disorder in an indigenous person, may well feel that however adequate European textbooks may have been for the purpose for which they were designed, the standard methods of examination given therein will now need supplementation from other sources. What follows is an attempt at partially filling the gap. (Burton-Bradley 1965b: 2)

The examination was institutional. He comments, 'Communication is difficult in the general medical ward, and it is therefore wise to use a private room during the process of examination' (ibid., 2). While Burton-Bradley was said to have travelled to most major towns in PNG at some time (Billig 1975: xii) there is no evidence that he ever worked outside an institutional clinical environment. Notwithstanding his huge literary output[33] on 'transcultural psychiatry' and his constant admonition that indigenous cultural factors should be taken into account in examination and diagnosis (see, for example, Burton-Bradley 1965b: 2–7), the procedure for treatment during his tenure simply involved hospitalization, ECT and drug treatment—the mainstream practice of European psychiatry of the 1960s.

Burton-Bradley was the only European psychiatrist in PNG for most of his sixteen years in office. During research in Port Moresby in 1985 I found a substantial body of anecdotes circulating among the medical community about his professional isolationism. The main theme of these was that he had gone to great lengths to keep other psychiatrists out of the country, and had made life so difficult for those who did come that they quickly left again. During our interview in 1985 Burton-Bradley offered with discernable relish an (unsolicited) anecdote about a psychiatrist who had come to PNG with his family and was posted to a rugged rural environment where he was unable to cope: he left the country after a few months. An exception to Burton-Bradley's alleged isolationism was his cooperation with Otto Billig of Vanderbilt University (Vanderbilt's press published Burton-Bradley's 1975 book *Stone Age Crisis*) and with E. D. Wittkower, a relationship which dated from Burton-Bradley's period in Malaysia (see Burton-Bradley 1973a: iii). A degree of territorialism is suggested by his documented response to the publication by Robin: His review of it in the *PNG Medical Journal* indicates that he took Robin's critique personally, and he tended toward *ad hominem* argument:

It has been my experience in the last twenty years to see a parade of ambitious young researchers in PNG whose primary purpose would seem to have been to acquire a sufficiency of superficial data to enable them to present a master's or doctoral thesis at some foreign university. Such an intention relieves them of the obligation to either construct policies and provide in-depth services or deal with the exigencies and economic limitations of operating a health service within a rapidly developing and changing environment. Relieved of responsibility in this way they are free to offer gratuitous criticisms and depart hurriedly to the academia of their choice where they are no longer obliged to defend their interpretations of hastily garnered selections of data in the face of informed judgement. (Burton-Bradley 1979: 203)

He also commented unfavourably on the advent in the West of 'the pseudo-humanitarian movement to empty all back wards that has placed society at the mercy of maniacally inclined psychopaths' (ibid., 205). When I interviewed him in 1985 he displayed some antipathy toward overseas agencies, particularly the World Health Organization (WHO).[34]

During the 1960s and 1970s the WHO and others were disseminating the concept of community health projects in the so-called 'Third World', emphasizing the development of primary health care, including mental health care, outside hospital settings. Nothing of this kind was seriously discussed with regard to mental health in PNG until after Independence. The priority announced in the Papua Annual Report 'to direct attention to members of the community who may be able to play the role of "community caretakers" in community health services' (TPAR 1967: 79) remained a paper resolution. In his 1973 publication *Longlong* Burton-Bradley made a gesture toward the idea of a service outside the hospital walls, involving a 'mental health educator' who would 'go into the village, sit down with the people and discuss their mental health problems' (Burton-Bradley 1973a: 257), but it is clear that he regarded this phase as subsidiary to hospitalization:

> He should spend most of his time in the field, and endeavour to influence attitudes to mental illness, to reduce hostility to the mentally sick and to suggest practical measures of management *at those stages in the history of the disease or maladaptive process that take place in the village.* (ibid., emphasis added)

In a reference to 'community contributions to service' he cautioned against laying 'too much emphasis upon subtleties to the neglect and exclusion of essentials' and opined that '[t]he invention of false problems, and uselessly complicating and prolonged debate on minor considerations are matters more appropriate to other walks of life than to those of the medicopolitics of social change' (ibid., 256).

Burton-Bradley's views were not entirely shared by his successor of 1975, Wilfred Moi,[35] and the administrative attitude toward 'primary prevention' and community agencies changed. Moi absorbed the clinical model of his teacher, but during the 1960s and 1970s he travelled overseas under the auspices of a number of international agencies including the WHO[36] and this perhaps accounts for his adoption of concepts disseminated by them. In an address to a World Psychiatric Association conference in Tahiti in 1976 Moi referred to the need to train a wider range of medical and para-medical workers and community leaders with the aim of 'primary prevention' of mental illness (Moi 1976: 3). He went on to comment, in a smooth articulation of the medical model and preventive work,

> In the day to day practice of medicine, we know that almost all physical pathology produces psychological reactions which in themselves threaten the individual's well be-

ing. In this sense, all preventive medicine has an inevitable component of preventative psychiatry, and every physician or healthworker, regardless of how narrow the concept of his speciality, when he is working in the field of social and preventive medicine he is working in the field of preventive psychiatry; he cannot avoid the role. (ibid., 6)

The drive to more community-based services preoccupied the mental health service after independence but was hampered by staffing problems, which were a legacy from the late colonial era. In 1972 the Cabinet of PNG's House of Assembly had approved a National Health Plan (which included mental health) for 1974 to 1978, concentrating on improved rural services, decentralization, self-reliance and simplification. The WHO commented in 1975 on the heavy reliance on expatriate staff in the upper levels of health services, and emphasized a need for post-graduate training for indigenous staff (WHO 1975: 16). In 1979 Robin wrote that general nurses selected for mental health service received an apparently comprehensive one-year course at the Laloki psychiatric centre, but said its effectiveness was limited by its being hospital based, and students had 'little or no opportunity to apply their training within a rural, village or settlement environment. Lack of financial support, transport and emphasis by higher level mental health authorities make field work almost non-existant' (1979: 46). He pointed to a lack of training of psychiatric staff for more responsible jobs, adding that mental health charge nurses operated in isolation and received little support or advice (ibid., 47). He further claimed that there was a notable departure of nurses from the mental health service during the 1970s, citing as reasons frustration at confinement to ward duties, lack of personal assistance or interest by higher level mental health staff[37] and constant exposure to pathological behaviour of patients (ibid., 45).

Colonial officers' inadequate training of, or provision for, replacement staff created problems in many areas of administration after the European exodus at independence. In the mental health area, any policies for extension of services were severely hampered by a lack of trained staff. A WHO advisor who visited in 1978 was quoted (with concurrence) by Robin the following year as writing in an unpublished report that mental health services in PNG were still fundamentally hospital-based rather than community-based, there was little opportunity for preventive work, and inefficient follow-up and after-care of discharged patients. The only way to improve the situation, said the WHO representative, would be a decentralization of mental health services to the community level and intensified training of ground-level staff (Robin 1979: 56).

Despite WHO involvement in the late 1970s in the revision of the post-basic psychiatric nursing programme to include community-based approaches (WHO 1981: 37), mental health services remained virtually unchanged from the completion of the Laloki hospital in 1967 to the mid 1980s. The aims expressed by Moi and the WHO had no practical manifestation, and Burton-Bradley's withdrawal from active involvement, coupled with Moi's withdrawal from practical work into the bureaucracy, created a gap at the point of articulation between the bureaucracy and workers in the field – the superintendancy of Laloki hospital. Moi had been installed as medical superintendant of Laloki when it was opened, and from the time of his promotion to the assistant directorship in the health department the hospital was overseen by a variety of 'acting' heads from various backgrounds.[38] The discontinuity worked against any systematic dissemination of policy from the bureaucracy.

From the 1980s to the New Century

In the early 1980s the situation began to change with the emergence of an active mental health administrator, Br Andrew,[39] a trained psychiatrist working out of the Goroka hospital in the Eastern Highlands Province. Citing National Health Plan objectives to integrate mental health care within general health services and for all health workers, 'in particular physician specialists', to undertake responsibility for mental health services, he published a handbook for medical officers and administrators on the management of psychiatric disorders (Andrew 1983). It was the first of its kind in the country. Incorporating sections on diagnosis, admitting procedures, the administration of ECT, drug treatment, and giving evidence in court, the handbook was a reaffirmation of the medical model at its simplest, a ready-reference for the non-specialist. Its nature had largely been determined by the current state of mental health services. A psychiatric hospital, and psychiatric wards at all the main hospitals, had been established but the country lacked the attendant mental health specialists. General medical practitioners at the hospitals were occasionally presented with madpeople whom they felt unqualified to deal with, and the handbook solved their problems by providing a guide to diagnosis, a guide to using the ECT machines which had been installed during the 1960s, a guide to drug régimes and a guide to the forensic aspects of psychiatry. The efficacy of Andrew's publication is hard to gauge, however, because staff transfers were common and the bulk of medical officers were still Europeans by the mid 1980s, on contracts and thus ever-changing. Most doctors I spoke with in Port Moresby and the highlands in 1985 and 1986 were unaware of the book's existence.

By 1985 Br Andrew had become 'Senior Specialist Medical Officer, Mental Health', a position brought into existence during bureaucratic shuffling and re-definition of jobs during the early 1980s. He was concerned with developing a primary prevention system to supplement the hospital system. To this end he overcame the lack of personnel by involving an organization called the Christian Institute of Counselling (CIC). The highlands-based group evolved in 1984 from a counsellor training programme begun in 1981, an ecumenical effort to transpose basic concepts from Western counselling practice to a PNG context. It involved finding community-minded indigenous people in urban and rural societies, training them in 'listening skills' and utilizing their local cultural knowledge to understand and solve problems of distressed individuals. The organization claimed to have significant success, especially in cases of depression, via simple catharsis.[40] In Goroka, where it was based, the European doctor temporarily in charge of the regional hospital's psychiatric ward in the mid 1980s was extensively using its services as an alternative to treating patients in the conventional manner.[41] At that time the CIC was embarking on a regional expansion of its training programme and attempting to interest village court officers[42] in learning 'listening skills'; and a number of church organizations were contributing with professional counselling skills learnt during their missionary training in the West (CIC 1985).

However, the status of the CIC scheme did not change from that of an adjunct of the mental health service to that of a substantial constituent. Br Andrew, the most influential figure in the service in the 1980s, shared the theoretical direction of Sinclair in the 1950s – a medical model and a 'stress' theory orientation (Andrew 1983, 1985; and see chapter 2, this volume). The psychiatric ward or hospital remained the ultimate repository of the mad. As there were no roads from the highlands to Port Moresby, transport problems determined that Laloki's geographical catchment area tended to be Papua, while Goroka hospital's psychiatric ward received highlands patients. The Goroka ward was 'secure' (i.e. it had lockable rooms), compared to the psychiatric sections of other highland hospitals, rendering it the repository of the more florid cases in the highlands region.

At Laloki in the mid 1980s, when I visited before embarking on my fieldwork, the complex was still as it had been in the 1970s, and as Robin had described it in his critique (1979: 48–52), though work had started on a new building, with three wings radiating from a central point, to replace the previously scattered wards. The existing complex was externally pleasant; small buildings were separated by stretches of grass and flowering shrubs and there was a notable lack of fencing, giving it an open and relaxed aspect. The only clue to its function was a sun-worn painted sign where the grounds met an unsealed road about 15 kilometres from Port Moresby. Male and female patients were housed mainly in two open, undecorated wards with security rooms for emergencies, such as violent outbursts. Some distance away was the 'rehabilitation village' consisting of a ward known as the 'Burton-Bradley Ward' and a garden area. The most vivid memory I have retained of Laloki from that period is of the 'Kokoda Ward', the maximum-security ward housing dangerous patients and 'His Majesty's Pleasure' patients. The latter patients had been there for years, forgotten by the incarcerating authority and trapped by a legality which required a government representative's intervention to secure their release. The ward was dark and its heavy clanging doors and tattered mattresses contributed to an atmosphere of sinister melancholy redolent of old-fashioned dungeons. Attached to it was a barbed-wire fenced outdoor enclosure. In 1985 a temporary 'acting superintendent' was humanely attempting to empty the ward but was legally unable to free the 'His Majesty's Pleasure' inmates.

While the Kokoda Ward ostensibly represented the dark side of psychiatry – the remnant of the asylum – its contrast with the open rehabilitation village becomes less distinct when the real social position of the village patients, allegedly being prepared for a return to society, is considered. In his 1979 description of Laloki, Robin pointed out that the rehabilitation ward, 'the showpiece of the entire centre' with its village atmosphere, was in fact populated entirely by males, some of whom had been away from their villages for up to twenty years and few of whom were in contact with their families or clan groups (1979: 51–52). This was still the case in the mid 1980s. The acting superintendant of the time told me the possibility of complete rehabilitation was severely hampered by the lengthy separation of the patients from their homes.[43] In this respect the new brick building, completed in the late 1980s, with its three wards radiating from a central control point brought the appearance of the institution together with its social reality.

By 1990, when I returned to PNG, there was a new head of mental health services, a Papua New Guinean trained in general health services rather than in psychiatry. His expressed desire to develop extramural care for the mentally ill[44] suggested that emphasis would be placed on the development of community mental health care, but no significant developments have occurred between then and the time of writing. While portrayals of PNG as a 'failing' state (see, for example Hughes 2004) have been rightly criticized for their superficiality (see, for example, Sullivan 2004), few would deny the country's institutional stagnation (see, for example, May 2001). Medical services, including psychiatric services, suffer like many other elements of state from the lack of implementation of successive policy declarations, frequent staff changes at government department level, and inefficiency or 'corruption' in the delivery of resources (see, for example, article titled 'Health Fraud is "Rife"', *Post-Courier*, 20 June 2007). Acting heads of psychiatric services have come and gone with regularity, consigned by their temporary status to the maintenance of a status quo, rather than the practical development of the visions of the late 1970s. The regional hospitals continue in the twenty-first century to have nominal psychiatric wards, with a very limited range of psychiatric drugs (chlorpromazine is still the most widely used medication) but they are chronically understaffed, usually by people with at best a minimal psychiatric nursing training, and on the few occasions I have visited (albeit very briefly) in the 1990s and into the 2000s they have been barely functional. In Port Moresby, the general hospital retains a psychiatric ward ('Ward Six') with fifteen beds, and I have discerned no changes in recent years worth remarking upon. The architectural change at Laloki in the 1980s has not been followed by any perceivable development in psychiatric practice. The Medical School at the

University of PNG continues, at the beginning of the twenty-first century, to try
to provide training progammes for doctors, with support from foreign quarters,
including a Masters in Psychiatry programme operated in liaison with the New
South Wales Institute of Psychiatry in Sydney, Australia. A handful of students
have completed the program.

Commentary

If the history of psychiatry in PNG were reduced to a chronicle of events the
reader might infer progress, or at least its intent, until the late 1980s. If the political context were ignored, for instance, we could conceivably be drawn into the
central and immediate movement recorded in the incomplete health department
files – the struggle to develop adequate facilities – and later the need for more
trained staff and the dispersal of mental health care into the community which
concerned commentators like Robin and the WHO in the late 1970s. Further, the
Sinclair Report and the creation of a Division of Mental Health in the late 1950s
present themselves as the threshold between the prehistory and history of psychiatry in PNG. Such a view is implicit in Burton-Bradley's remarks that psychiatry
was born in a galvanized iron shed at Bomana. This sense of progress is enhanced
by the view that orthodox clinical psychiatry had of itself in Burton-Bradley's day.
It was represented as a speculative science whose special province was 'the understanding of those disorders of subjective experience or objective behaviour which
are themselves a cause of disability' (Stafford-Clarke 1964: 15), and whose clinical practice was 'still more of an art than a science' (Solomon and Patch 1974:
3). Historically, its practitioners had come to see it as a humanitarian science;
for the replacement of the asylum with a hospital and the implicit reappraisal of
the bestial lunatic as a sick human was the historical horizon which psychiatry
had set for itself. William Tuke's establishment of the York Retreat in 1796[45] and
the questionable (see Midelfort 1980: 258) story of Pinel liberating the mad of
Bicêtre asylum from their chains are the stuff of psychiatry's genesis myth:

> [When Pinel] removed the chains from the patients under his care, he released by this
> act not simply the bodies of the patients, but ultimately the minds of his colleagues,
> from preoccupation with cruelty and fear as permissible ways of treating the mentally
> ill. (Stafford-Clarke 1964: 20)

Since conventional psychiatry defines itself less by substance than by these
two points of departure – one historical, one intra-medical – it is not surprising
that the Sinclair Report and the decision to replace the Bomana asylum with
the Laloki hospital seem to be psychiatric turning points. Yet the paraphernalia
of psychiatric practice (ECT, drugs) were already being used in PNG as early
as 1950 and if Kettle's memoir is correct lobectomies were performed in the

late 1940s. Further, the treatment of the mad, using drugs and shock therapy, did not change significantly after the establishment of the mental health service. This chapter has indicated that historical changes were in fact bureaucratic and architectural, and directly linked to ideological factors rather than to any mental health needs of Papua New Guineans. The Bomana asylum reflected, with crude honesty, the colonial attitude: It was a dumping place for unmanageable natives. This function took precedence over any question of care, the presence of the concerned A. J. Prior notwithstanding. The asylum was a response to the problems of white doctors and administrators, a place to send individuals who disrupted medical wards or gaols. Lobectomy, electroshock and tranquillizing drugs were answers to behavioural problems, and quite unrelated to 'the understanding of those disorders of subjective experience or objective behaviour which are themselves a cause of disability' (Stafford-Clarke 1964: 15).

The ostensible change in attitude (from control to care) represented by the Sinclair Report and its recommendations for a systematic mental health service was not based on evidence of any significant occurrence of madness requiring attention. Sinclair's own comparison of the contemporary admission rate to Bomana asylum with that to hospitals in Queensland, Australia – 0.22 per 10,000 in PNG compared to 8.5 in Queensland (Sinclair 1957: 30) – revealed a negligible observable incidence of madness that hardly required the creation of a special service. In addition, there is a contrast between the report's theorizing on stress and psychoses (ibid., 30–32; see chapter 2, this volume) and a comment that the requirement for a mental hospital was related to urbanization bringing with it 'the expected *lowering of tolerance* for mental illness in native society' (Sinclair 1957: 29, emphasis added). The latter statement implicitly acknowledges sociopolitical factors in psychiatric hospitalization, and the low statistical incidence of hospitalization reported by Sinclair does not seem particularly worthy of the 'urgent attention' response referred to by Hasluck. In the prevailing climate of post-war colonialism a brute response to the expectation of less tolerance of madness would have been to build more asylums or enlarge the existing ones; and inasmuch as treatment methods remained unchanged after the establishment of a 'mental health service', this is arguably what happened. The barbed-wire fence and sheds of Bomana were replaced by the open-plan buildings of Laloki and the separated psychiatric wards of the regional hospitals. The lunatics and mentally deranged natives became mentally ill patients; nevertheless, they were sequestered and given ECT and tranquillizers just as before.

I observed earlier that the Sinclair Report brought to the social Darwinism of the colonial régime the conventional wisdom of psychiatry of the time, a theoretical approach combining clinical and 'stress' factors. This, too, had social Darwinistic precedents: In the nineteenth century, when the stresses of civilization were first mooted as a cause of madness, the obverse image was the noble savage, an image grounded in fancy and traveller's tales alleging that madness was never seen in primitives in their natural and simple environment (Mellett 1982: 66–70). The theoretical input of the Sinclair Report is important in light of the colonial Administration's concern with 'development', for it provided a justification for the continuation of the internment of the mad. Previously the 'criminally insane' or 'mentally deranged' natives had been put into the asylum because they were (as the descriptive language implies) uncontrollable. Under the new policies such attitudes were unacceptable. Concern for the wellbeing of indigenes under the onslaught of civilization became the new justification. Viewed from this perspective the mental health service presaged by the Sinclair Report no longer appears as a liberal and progressive force but as a superstructural element largely determined by the acceleration of 'development' in the late colonial period.

Following the establishment of a formal mental health service, the rift between its ideological development and the practical treatment of the mad became increasingly pronounced. The Mental Disorders Ordinance of 1960 continued an illusion of reform and progress. It devoted several pages to a juridical pronouncement on the management of patients' business interests and estates, in a country where principles of kin-group organization dominate property relations, and it introduced criteria for hospital visitors, against social and geographical factors which determined, as Robin pointed out, that psychiatric patients would often be irreconcilably separated from their kin groups. At the same time it consolidated the existing real conditions of treatment by providing a regulation enabling a wider use of ECT.

Burton-Bradley, while producing literature emphasizing the risks of misdiagnosis in a cross-cultural environment, trained psychiatric nursing staff in standard practice and was resistant to trends 'which may have transferred the control of madness back into the community. Considered in light of the community mental health orientation of the immediate post-colonial period, Burton-Bradley, an idiosyncratic and conservative thinker, might appear as a recalcitrant figure in the drive to 'progressive' mental health care, but it should be remembered that his role was ideologically circumscribed insofar as he was installed to implement the import of the Sinclair Report. The fact that the Laloki institution, a very simple project involving the construction of half a dozen light wooden buildings, took eight years and a great deal of correspondence to complete, suggests that Burton-Bradley, together with Prior and Scragg, was attempting to implement something that the Australian Ministry for Territories regarded as window dressing. In this respect the Papua Annual Report's reference to adequate and improving psychiatric facilities (when they were barely existent) and the Controller of Corrective Institutions' warning about damage to the Administrations reputation from press revelations of gaol-held mentally ill patients are especially salient. Having struggled to install an institution in keeping with the psychiatric principles under which he was appointed and to which he subscribed, it is not surprising that

Burton-Bradley was resistant to trends which he perceived as diverging from the clinical system.

The post-colonial preoccupation with primary prevention and community-based services has a double aspect. Br Andrew's publications (1983, 1985) make it clear that although he was more disposed to the concept of primary prevention than Burton-Bradley appeared to be, he distinguished between a degree of distress which trained counsellors could help and more complex forms of madness requiring hospitalization and clinical treatment. This view is institutionalized in Western societies in the duality of psychotherapy and psychiatry, with the former involving a contractual relationship between a counsellor/therapist and a 'client', and the latter involving committal or submission to medical-scientific treatment. In the case of PNG it maintained a need for psychiatric institutions and perpetuated the status quo. The concept of 'community-based' services lacks specificity, but as it usually involves a system of dispensing antipsychotic medication (while excluding ECT automatically by virtue of being outside the hospital) it invariably becomes, as Anthony Clare once pointed out, 'a delivery system that organizes psychiatric facilities in a particular way' (1976: 55).

Whatever concerns may have been expressed about developing community-based treatment, the social reality was demonstrated by the new facility built at Laloki in the 1980s. The replacement of the open-plan wooden buildings by a star-shaped brick structure controlled from its central point had a historical irony. In England in 1791 Jeremy Bentham proposed a new design for asylums:

> From a central room permitting an overview, corridors running along workrooms or cells are to radiate out as from a star to facilitate surveillance by a single person, or two or three at most – i.e., maximum efficiency at minimum cost ... Bentham saw his plan as a gesture of liberalization for the insane: chains and other such physical restraints were supplanted by a more efficient architectonic-organization restraint. (Doerner 1981: 72)

Socially, then, the locus of change in the treatment of madness in PNG was not the coming of psychiatry in the late 1950s, but the intensification of control immediately following the Second World War, when sequestration of the mad by European administrators began. In its material effect on madpeople, the shift in the colonial attitude from pre-war disinterest to control is more salient than the later shift from control to 'care', the latter having left the fate of the mad unchanged. The continuation of pre-psychiatric treatment – shock and drug treatment – compared with the various and changing expressed concerns of psychiatry, such as 'care', 'rehabilitation' and 'primary prevention', suggests that psychiatry has functioned on two separate and contradictory levels: one practical, perpetuating a previously existing form of social control, and the other theoretical, expressing concern for mental health.

Notes

1. It needs to be said that Monckton was prone to exaggeration and his self-serving anecdotes are often of questionable veracity. They do, however, offer valuable insights into the administrative attitudes of his time.
2. Monckton's non-intervention is perhaps a measure of callousness. His treatment of Papuans was relatively brutal compared to most of his fellow officers.
3. National Archives of Papua New Guinea, Port Moresby (hereinafter NAPNG), Dept. of Health Registry File No. 18-13-4, 'Welfare Natives, Mentally Deranged'.
4. NAPNG, Dept. of Health Registry File No. 18-13-4, 'Welfare Natives, Mentally Deranged'.
5. NAPNG, Dept. of Health Registry File No. 6-3-3, 'Buildings, Mental Asylum'.
6. NAPNG, Dept. of Health Registry File No. 18-13-4.
7. NAPNG, Dept. of Health Registry File No. 18-13-4.
8. NAPNG, Dept. of Health Registry File No. 6-3-3.
9. Interviewed by author, 24 September 1985.
10. Etai Tiamu, interviewed by author, 30 September 1985.
11. Burton-Bradley repeated the 'rats' story when I interviewed him in 1985. The story has an odd precedent in notes by Dr D. F. Macgregor, a clinical psychiatrist then working in Fiji, of a meeting between himself, Burton-Bradley and Dr G. Zeldine, a neuropsychiatrist working in New Caledonia. Macgregor notes that after hearing a report by himself on urbanization and mental illness in Fiji based on long-term statistics, Zeldine expressed regret that he could not produce the same kind of data, explaining that records of his New Caledonia hospital had been eaten by rats. The notes record no comment by Burton-Bradley (Macgregor 1967: 6).
12. NAPNG, Dept. of Health Registry File No. 6-3-3, correspondence, 15 September 1954 and 30 November 1954.
13. Ibid., correspondence, 2 September 1957.
14. Ibid., 1960.
15. After the Second World War, Australia administered Papua as its own territory and New Guinea as a United Nations Trust Territory. While the country was a single unit as far as practical administration was concerned, the Trust Territory Annual Report was a separate document forwarded to the UN. While the Papua report and the New Guinea report were identical in many respects, the relationship between Australia and the UN was reflected in the occasional modifications and explanations which appeared in the New Guinea report.
16. Burton-Bradley, pers. comm., September 1985.
17. NAPNG, Dept. of Health Registry File No. 6-3-17 'Buildings and Hospitals, Laloki'.
18. Contents of a letter from Scragg to Burton-Bradley (at an Australian address) dated 23 December 1960 implied that Scragg, at least, was at the time conceiving the Laloki project as only a mental hospital (NAPNG, Dept. of Health Registry File No. 6-3-17).
19. NAPNG, Dept. of Health Registry File No. 6-3-17.
20. Fragments in Health Department files suggest recriminations between that department and the Survey Department, which allegedly knew of the river's propensity to flood at that location but had omitted to impart that information during planning (NAPNG, Dept. of Health Registry File No. 6-3-17).

21. Letters from Burton-Bradley recommending the issue of certificates were found in Archive files; NAPNG, Dept. of Health Registry File No. 25-1-10, 'Mental Disorders and Treatment Ordinance'.

22. In our 1985 interview, Burton-Bradley expressed suspicions to me that funds had been 'diverted' to other projects.

23. NAPNG, Dept. of Health Registry File No. 6-3-17, letter from A. J. Prior, 17 June 1964 *passim.*

24. NAPNG, Dept. of Health Registry File No. 6-3-17, letter, 18 June 1963.

25. *South Pacific Post*, 23 December. 1964.

26. NAPNG, Dept. of Health Registry File No. 6-3-17, letter, 24 December 1964.

27. NAPNG, Dept. of Health Registry File No. 6-3-17, letter, 1 April 1965.

28. Ibid., letter, 2 April 1965.

29. NAPNG, Dept. of Health Registry File No. 6-3-17.

30. NAPNG, Dept. of Health Registry File No 25-1-10.

31. NAPNG, Dept. of Health Registry File No 25-1-10, 4 March 1964.

32. This was the first report to the UN combining Papua and New Guinea into a single administrative entity.

33. During our interview in 1985 he claimed more than 100 publications. It has to be said that there is a considerable amount of repetition among his articles.

34. Having implied that Burton-Bradley was isolationist and territorial I should in fairness report that when I met him he was courteous and accommodating, and gave me valuable historical information in a long interview without reservations.

35. After his replacement as the national head of psychiatric services, Burton-Bradley retained administrative control over the psychiatric ward at Port Moresby General Hospital until the late 1980s, and remained on the editorial board of the *PNG Medical Journal.* His relationship with Laloki was 'phased out' during the decade following independence. He continued to practise psychiatry at his home in Port Moresby. He received a knighthood in 1990 and died of a heart attack in 1994.

36. W. Moi, pers. comm., 1985, and Burton-Bradley, pers. comm., 1985. Burton-Bradley opined that Moi had been seduced by the WHO and he clearly regarded this as a bad turn for psychiatry.

37. In 1985 I heard similar (retrospective) comments about lack of interest in relation to the psychiatric ward at Port Moresby General Hospital in the 1970s.

38. Wilfred Moi was no longer actively engaged with bureaucracy when I interviewed him in 1985. He was running a community health clinic in Port Moresby. He died in 1995, from stab wounds received when he intervened in a bag-snatching attempt outside a Port Moresby restaurant.

39. English Franciscan.

40. Sister Judith de Montfort, SMSM, director of CIC, pers. comm., 1986.

41. J. Richens, pers. comm., December 1985.

42. The 'village court system' was legislatively introduced. The first village courts began operating in 1974 as a grassroots dispute-settling service. Village court 'magistrates' were untrained in law and were intended to apply 'customary' judicial wisdom in their practice (see Goddard 2009).

43. R. Trompf, pers. comm., 1985.

44. K. Chakumai, Assistant Secretary, Mental Health Services, pers. comm., July 1990.

45. At a pastoral site with open grounds, Tuke attempted to follow a philosophy of treating insanity, offering patients the opportunity of quiet reflection and appealing to their moral sensibilities and sense of worth (see Tuke 1964).

2

PSYCHIATRIC THEORY AND PRACTICE IN PAPUA NEW GUINEA

Psychiatry seeks to dim the deeply disturbing and uncontrollable phenomena of nature and render nature itself harmless, so that what remains can be dealt with theoretically within a harmonious, humane, and scientific framework. A grandiose process of camouflage idealizes nature. Praxis is limited to hard-to-verify success in individual cases, or becomes purely administrative.

—Klaus Doerner, *Madmen and the Bourgeoisie*

In the previous chapter I contextualized psychiatry in PNG in the history of colonialism and its aftermath. From that perspective its contribution as an aspect of social control has been foregrounded, implicitly challenging psychiatry's scientific and therapeutic self-understanding. Precedents for the perspective I have adopted can be found in a variety of critiques of psychiatry in Western society which have emerged since the 1960s. In one way or another, they have positioned psychiatry as a historical product of specific social, political, economic and ideational developments in Western societies (see, for example, Basaglia 1980; Doerner 1981; Foucault 1982, 2006; Mellett 1982; Rosen 1968; Scull 1979, 1981). A corollary of this body of critique is the observation that psychiatric diagnosis and treatment in Western societies is determined to a significant degree by contemporary social, political and economic circumstances which are rarely acknowledged in the diagnosis and treatment itself.[1] There has been, in other words, an inherent contradiction between psychiatric theory (the project of therapeutic treatment of mental illness) and practice (the containment and control of socially disruptive individuals). Analysing the process by which psychiatry in PNG was institutionalized during changing colonial administrative attitudes and policies, I find this contradiction to have been perpetuated under colonial rule and to have persisted post-colonially.

In this chapter I investigate the schism between psychiatric theory and practice in PNG with particular regard to ideas of 'transcultural' or 'cross-cultural' psychiatry. Both of these adjectives imply the possibility of operational modifications to psychiatry in response to a non-Western social context. Here I critique psychiatry's putative operational sensitivity to 'cultural' difference with clinical examples from the colonial and post-colonial eras, up until the mid 1980s. It hardly needs to be said nowadays that the attempt to apply psychiatric ideas about madness to non-Western societies requires at least a consideration of local 'cultural' factors and contexts, as psychiatrists and anthropologists have acknowledged since the mid twentieth century.[2] Indeed, transcultural psychiatry has undergone some modifications, in theory at least, since I conducted my fieldwork in the mid 1980s. For example I suggested in a paper in 1991 that *laissez faire* psychiatric perspectives allowing for the existence of 'culture-bound syndromes' (which therefore may require no psychiatric intervention) might inadequately conceive both the 'culture' and the 'syndrome' (Goddard 1991). It could now be argued that my comment has been obviated by transcultural psychiatric self-critique, for recent debate implies a recognition of the inadequacy of the concept of culture to which psychiatry once subscribed. A significant body of transcultural psychiatrists have now moved beyond a view of culture as a bounded, homogenous entity, and prefer to problematise its hybridity, especially in relation to discourses of the engagement of the global and the local (see, for example, Kirmayer 2006; Miyaji 2002; Waldram 2006). However, brief reflections on recent anthropological discussions of culture do not encourage confidence that transcultural psychiatric discussion will achieve universal *verstehen* early in the new century.

The idea of culture has become increasingly slippery for anthropologists through a variety of contextual developments. As well as the imperative to recognize 'global' influences and hybridity, there have been calls to counter the contemporary dissemination of 'essentialized and racialized views of culture(s)' (Trouillot 2002: 38), and to salvage its analytic validity from 'vapid' and 'insipid' (Fox and King 2002: 2) usage in popular expressions. Preoccupations such as these add to the difficulties manifest in successive attempts at analytically useful definition by anthropologists stretching back at least to a mid-twentieth century survey by Kroeber and Kluckhohn which catalogued the dozens of overlapping but different definitions (Kroeber and Kluckhohn 1952) which had emerged by that time. Since the 1980s there have been a number of negative assessments of the analytic value of the culture concept from differing perspectives. For example, on behalf of people such as feminists and 'halfies' who found themselves excluded by common views of 'culture', Abu-Lughod suggested that 'anthropologists should now pursue … a variety of strategies for writing *against* culture' (1991: 138, emphasis in original). Alternatively Kuper, citing among other things the oppressive political potency of popular concepts of culture, observed that 'if the elements of a culture are disaggregated, it is usually not difficult to show that the parts are separately tied to specific administrative arrangements, economic pressures, biological constraints, and so forth' (1999: 246) and argued that 'unless we separate out the various processes that are lumped together under the heading of culture, and then look beyond the field of culture to other processes, we will not get far in understanding any of it' (ibid., 247).

Alternatively, those who wish to preserve the concept despite the ravages of popular and political appropriations have sought to recast it conceptually. For example Christopher Hann thinks it is time to 'demote culture from its quasi-mystical status as the discipline's master concept ... [C]ulture is best understood as *congealed sociality*, a transient patterning of clusters of behaviors and ideas' (2002: 273, emphasis in original). Another strategy has been to adopt an adverbial rather than nominative use of the concept, given that 'culture is not something we can ever expect to encounter "on the ground"' (Ingold 2002: 330). Tim Ingold suggests 'It might be more realistic, then, to say that people *live culturally* rather than that they *live in cultures*' (2002: 330, emphasis in original). Responding to the general critique, Sahlins has chastised critics for unfairly representing early modern American anthropologists as constructing a notion of cultures as 'rigidly bounded, separated, unchanging, coherent, uniform, totalized and systematic' (1999: 404), which they subsequently critique with appeals to global influences and hybridity. For good measure, he takes arms against anthropologists who 'traded in their cultural heritage for a mess of Foucauldian discourse, or the mess that has been made of Foucauldian discourse' (ibid., 410). From the perspective of his own variant of structuralism Sahlins argues that the 'dialectic of similarity and difference, of convergence of contents and divergence of schemes, is a normal mode of cultural production' (ibid., 411). Whatever stance one might take on the problem of 'culture', an implication of its increasing problematization among anthropological discussants is that psychiatry's attempt to dilute its own ethnocentrism with any degree of cultural relativism may be in vain.

Transcultural Psychiatric Theory in PNG

I take as my historical starting point the Sinclair Report of 1957 which I discussed in the previous chapter in terms of its practical impact. Its significance in the present chapter is that it contained the first theoretical statement of direct relevance to psychiatric practice in PNG. Sinclair subscribed to a conventional 'stress' theory of psychosis which we should articulate here since it has been a recurrent theme in psychiatry in the country ever since. Sinclair distinguished between a neurotic and psychotic reaction to stress, according to the nature and intensity of the stress and the character of the individual (1957: 30). The neurotic defence, he said, was usually regarded as a less complex stage in the defence mechanism against a stressful environment. 'The neurotic retains contact with his environment. In so doing he employs physical symptoms as a means of retaining emotional contact with individuals within the environment. The presence of neurotic disability also allows him to avoid situations too stressful for him to tolerate' (ibid.). By contrast the deeper, psychotic reaction involved a withdrawal from the environment rather than a manipulation of it. The deeper-seated stresses, precipitating psychosis, arose from 'long-standing conflicts between instinctual urges and striving on the one hand, and the controlling influence of training, social pressure, precent and taboo in [*sic*] the other' (ibid.).

Sinclair then posed two questions: Was the pattern of psychosis changing, and would increased urbanization influence the pattern? (ibid., 31). His attempt to answer the first question was curious insofar as his was the first psychiatric survey in the country and had no diachronic data to suggest patterns in the first place. The answer itself was confused and appealed to aggression as a gauge:

> The behavioural pattern in psychosis is certainly changing. This does not mean that causative factors are changing. The most remarkable change in the pattern of psychosis is the reduction in the amount of aggressive or violent hysterical behaviour. In a general way our inquiries suggest that this reduction has gone *pari passu* with reduction in inter-tribal fighting consequent upon Administration sanction against fighting. It is a well-known fact that the pattern of hysteric expression is determined by custom and fashion and it may well be that aggression and violence have lost their 'fashionableness' both in the behaviour of normal and psychotic natives. When the problem of causation in the long-standing psychoses is considered, it is safe to assume that in these more permanent withdrawals from reality the causation is not likely to be influenced by the relatively superficial change in cultural pattern. (ibid.)

Sinclair's conflation of 'customary' hysteria and psychosis was the prelude to a theme which was to become central both as the 'problem' and as the theoretical fuel of psychiatry in PNG, as we shall see later. The answer to the second question, concerning the influence of urbanization, was less confused: '…[I]t is likely that the native, confronted with increasing conflict, is more likely to break down at the first line of defence rather than at the second. In other words, he is more likely to become "neurotic" than "psychotic"' (ibid.). This speculation was followed by another that 'should the problems of interpersonal and social organization reach the complexity which obtains in European culture, *then* the incidence of psychosis … will increase to the ominous rate at present obtaining in our own culture' (ibid., emphasis in original).

It is worth noting that Sinclair's initial prognosis was mild; he anticipated neurosis rather than psychosis – hardly a justification for psychiatry – and his afterthought can be reduced *ad absurdum* to the proposition that if PNG society were structurally like contemporary Western society the incidence of madness would be similar. As an argument for a mental health service in PNG society at the time, the Sinclair Report had no theoretical strength, and axiomatically it reproduced an old fallacy, for implicit in Sinclair's speculations was a lack of stress in 'traditional' society, at least compared to Western society. This assumption has pervaded transcultural psychiatry from its beginnings, without being seriously questioned until its supercession at the beginning of the twenty-first century by a new preoccupation with the problems of defining culture (see above). The power of the fallacy of unstressed traditional society was evident for example in a 1983 WHO report on depressive disorders which stated in its opening paragraph that:

> [P]eople are now living in a rapidly changing social and physical environment that often gives rise to acute or prolonged psychosocial stress and may lead to depressive reactions. At a time when traditional protective mechanisms of social groups are breaking down and many people are exposed to the unsettling effects of uprooting, family disintegration, and social isolation, the prevalence of depressive disorders arising as a response to stressful psychosocial factors … is likely to increase. (Sartorius 1983: 1)

The concept of the stress-free primitive was in part an unquestioned survival from nineteenth-century British psychiatry, when madness was attributed to the stresses of civilization and the British lower classes were regarded as the most susceptible, lacking the self-discipline and education to avoid psycho-moral degeneracy (Mellett 1982: 62–70; see also Doerner 1981: 146–47; Scull 1979: 228–30). Of course, the noble savage had an obverse role in this scheme of things (see chapter 1, this volume), but if he could be heuristically related to any section of the population of the civilized world, it was to the poor and uneducated. The inference was that primitives would be at risk of madness if exposed to civilization because like the lower classes of Britain (at best), they were intellectually ill-equipped to cope with its stresses.

But the pervasion of the image of the vulnerable savage in transcultural psychiatry also owed much to twentieth-century social anthropology, specifically the functionalist-derived ethnography which guided psychiatry's attempts to deal with 'culture' in non-Western societies. The heuristic methodology of early modern anthropology posited those societies as self-contained units whose equilibrium was maintained through the integration of their constituent social institutions. The resulting model was of static societies whose survival was conditioned by inner consistency and whose equilibrium was vulnerable to external, rather than internal, forces. While it did not necessarily reflect the everyday perceptions of the ethnographers,[3] functionalist ethnography gave 'primitive' communities a gloss of consensus and homogeneity that implied a stress-free existence (ritual madness was usually seen as having a safety-valve function: see for example Bateson 1972; Reay 1965). The resultant facile articulation of functionalism with psychiatry in the case of PNG can be seen, for example, in Sinclair's speculation that '[i]t is probable that disintegration of the native social structure will mean that the psychologically important developmental period of childhood will then occur within the much narrower confines of a tight family relationship and not within the less traumatic settings of a group community' (1957: 31).

While nothing has occurred in the five decades since the Sinclair Report to lend credence to the idea that Melanesians would become psychotic under the onslaught of civilization, it served for years as a professional *a priori* (see for example Burton-Bradley 1975: 8). Ironically Burton-Bradley, an advocate of the stress-of-civilization idea (as his 1975 book title, *Stone Age Crisis,* implies), provided statistics suggesting, in fact, a paucity of psychosis in the country:

> During the several decades of my experience in all Provinces of this country, the general and psychiatric hospitals have been concerned at any one time with less than 300 psychiatric patients in a total population of almost three million people. Not only is this percentage (0.01 per cent) a mere fraction of those in the care of mental health systems of Western countries, it is to be noted that a significant section of even these people are confined by specific direction of the National Court. (Burton-Bradley 1985: 25)

He explained this lack of patients with the statement that psychiatry was not a hospital-based system in PNG and that the 'overwhelming majority of psychiatric patients have little connection with hospitals and are treated on an extramural basis' (ibid.). This was a remarkable claim in the face of Burton-Bradley's own resistance to community-based psychiatry and of the ongoing difficulty (recorded in the previous chapter) in implementing such a system.

Given the history of nosological ambiguity in psychiatry (Doerner 1981; Mellett 1982; Pilgrim 2007; Zilboorg and Henry 1941) and the problem of defining 'civilization', the stress-of-civilization theory is untestable. Within the psychiatric paradigm any attempt to test the theory during its popularity would probably have had to resort to an examination of hospital admission statistics over a period of time, perhaps contextualized in a consideration of the degree of 'urbanization' (a tangible substitute for 'civilization', 'modernization', etc.). This was never attempted in PNG. In any case the degree of urbanization in relation to the total population since 1959 has been so slight, the rural catchment area of psychiatry so vast by comparison, and migration patterns in the country so varied, that any such exercise would have had even less credibility than its superficial conception.

The weakness of the theoretical justification for establishing psychiatry was quickly masked by more exotic theoretical discourse – the problems of cross-cultural diagnosis. This theme was elaborated especially by Burton-Bradley throughout the 1960s, with its most extensive articulation appearing in his 1973 publication *Longlong.* The development of this discourse was polymodal. One modality was a cultural-relativist concern, informed by anthropological data, acknowledging different perceptions of madness, the existence of what came to be called 'culture-bound syndromes' (see, for example, Lebra 1976; Simons and Hughes 1985), the efficacy of traditional healing beliefs and the need for diagnostic caution on the part of Western psychiatrists (Burton-Bradley and Julius 1965: 10–26; Burton-Bradley 1973a: 5–39). These considerations worked against the legitimacy of psychiatry on their own, but they were in tension with a second modality in the discourse which was concerned with the distinction between organic and non-organic psychoses in a cross-cultural setting (Sinclair 1957: 31–38; Burton-Bradley 1973a), and a third modality which was concerned with the psychiatrist's ability to recognize genuine psychosis in bizarre social instances (Burton-Bradley 1965a: 27–30; 1973a).

This discourse provided ideological complementarity to the themes of the Sinclair Report. Where the stress-of-civilization preoccupation had served to draw Melanesians into the reach of psychiatry, the 'cross-cultural' considerations projected psychiatry into Melanesian 'culture'. Any lack of substance which may have become apparent in the 'stone-age crisis' warning with the passage of time was compensated for by the preoccupation with distinguishing the 'real' from the 'cultural', the mental from the medical, and the psychotic from the psychodramatic. In the process the *a priori* stressless savage was transformed into a moderately stressed and occasionally psychotic one, with the prospect of deterioration through the ravages of acculturation (see Burton-Bradley 1973a: 243). A succinct integration of the various modalities of transcultural psychiatry in PNG was demonstrated in a comment on 'cargo cults' (presumably supplied by Burton-Bradley[4]) in a 1973 Administration report to the United Nations:

> A word of warning needs to be raised in respect of cargo cult activities. The popular view and the social science view that these activities are logical and to be expected within the frame-work of developing cultures in Papua New Guinea is probably correct. An unfortunate by-product of this is that individual cultists are thereby considered never to be mentally sick. This is not in accord with the facts, and clinical examination reveals that some are. Mentally sick people are just as liable to be found among cultists as elsewhere and are in need of treatment in the same way as others. Despite the viewpoint that denies the possibility of overt mental disorder among cultists, observation shows that the nature of these movements in fact often allows abnormal personalities to function in various roles within the movement without attracting attention to the fact that they are mentally sick. Recognition of this fact, and of the widespread existence of cargo thinking, assists the mental health services in determining a normal pattern of behaviour against which to judge the healthy adaptation of the individual. (RPNG 1973: 186)

While the 'cross-cultural' discourse, shifting the focus of attention away from urban trauma toward Melanesian culture, provided ideological legitimation for the use of psychiatry in the colonial control of quasi-political disruption, it also implied the possibility of liberal psychiatric intrusion in rural society. In practice the development of community psychiatry and counselling was delayed until after Independence due to administrative factors (see chapter 1, this volume). Only in the mid 1980s was a deliberate preventive aspect of psychiatry inaugurated, through utilization of a counselling service and counselling courses for community service workers (CIC 1985: 3–8).

The emphasis on cultural factors gave psychiatry a veneer of sensitivity. The inference of the on-paper distinctions between 'true' psychoses and culturally explicable behaviour was that psychiatry had the ability to ensure that no one who was not genuinely mad would be incarcerated or treated as such, and cultural factors would be weighed in favour of an individual's sanity. The practical validity of this approach was of course questionable, and probably no more trustworthy than the recourse of Sinclair in his 1957 survey: 'We found repeatedly that native observers could resolve our difficulties in separating delusional material from current belief, simply by asking them did they think the belief was an insane one or not' (33).

While Burton-Bradley (1973a) cited personal cases of putatively culturally sensitive psychiatry in action, there is no evidence that this theoretical perspective has affected standard psychiatric practice to any significant degree. As there have been only a few professional psychiatrists in PNG since the 1950s the professional encounter with madness has mostly involved law enforcement officers and paramedics in the first instance controlling socially disruptive behaviour rather than delusions. The patient consequently arrives at the psychiatric ward either sedated or socially inaccessible (aggressive or withdrawn). Under the circumstances, psychiatric theory is displaced by incarceration and the attempt to make the patient socially manageable. Some examples of the psychiatric encounters of people from my field-research area, the upper Kaugel Valley in the Western Highlands Province, illustrate common patterns of psychiatric incarceration. I will give an ethnographic account of the Kaugel people in chapter 3. The short description below will serve for the purposes of the discussion here.

Psychiatry and the Kaugel Valley

The floor of the upper Kaugel Valley is about 2,200 metres above sea level, and Mt Giluwe, which rises to an altitude of nearly 4,600 metres, forms its western side. The local district is now called Tambul, the name of an administration post at the head of the valley. Tambul is about 32 kilometres west of the main provincial centre, Mt Hagen. Christianity (mainly Catholic and Lutheran) was introduced in the 1950s, and a fundamentalist church ran a 'bible college' close to Tambul. When I began fieldwork in 1985 the journey from Mt Hagen to the valley took about three hours by truck. The main, limited source of cash during my fieldwork was out-migrant labouring by young men.

The administrative personnel at Tambul during my fieldwork period were headed by an assistant district commissioner (ADC). Law enforcement was managed locally by a fluctuating police contingent averaging three officers, complemented by a 'grassroots' system of village magistrates and peace officers recruited in consultation with the populace.[5] Tambul had a medical post with ward facilities and a small room equipped for emergency surgery, though it was hardly used; all but minor surgery cases were transferred to Mt Hagen General Hospital. The medical post (known as a health centre) had limited supplies of chlorpromazine in tablet and ampoule form, but had no other psychiatric resources, unless one includes anti-epileptic medication in that category. The health centre's patient records went no further back than the previous two or three years (like the ADC office, the health centre suffered chronic senior staff changes, to the detriment of systematic filing). Details of psychiatric interventions or admissions from the 1970s and early 1980s were not locally available, but the ultimate repository of the mad from the subdistrict was the 'secure' ward at Goroka Hospital, a little over 300 road-kilometres away, where I was able to find case files for a number of people from the Tambul area who had been admitted as psychiatric patients during the late colonial era.

One of the earliest cases in the files was a man who was admitted in the early 1970s and discharged after four weeks. A short entry stated that he had been a student at the bible school (an American fundamentalist mission) at Tambul for two years. He had fought on one occasion with two other students at a basketball game and subsequently went crazy (the file used the Tokpisin term 'longlong'), acting strangely, talking oddly, allegedly raping a girl and stealing a bicycle. He was taken by police to Mt Hagen and then transferred to Goroka Hospital, where he was admitted and given 100 mg of chlorpromazine six-hourly: In European practice this is a common maintenance dose (the maintenance dose ranges from 25 mg to 100 mg three times daily). He made one attempt to abscond, and was found at Goroka market, where buses to Mt Hagen departed from. Apart from that he was a quiet patient. He was eventually discharged without continued medication.[6]

I traced the man at Tambul in 1986 (the name recorded in the Goroka file was wrong – a common occurrence, I was to find) by which time he was mature and locally respected as a 'good Christian'. Apart from the outburst of 1971 his behaviour had always been exemplary. He verified for me that he had been to Goroka Hospital in the early 1970s. He said the crazy behaviour had come on at the end of the year, the time of his exams and graduation from bible school studies. The first sign, he said, had been uncharacteristic anger and assault on the basketball court. Shortly afterwards he had begun using obscene language pub-licly, and had gone to the girls' dormitory at night, trying to abduct the girls. He had not known what he was doing at the time, he said: It had been as though he was dreaming, and when people had recounted his actions later he had not believed them. He had stolen a bicycle, at which point he had been taken to Mt Hagen Hospital and sedated. From there he had been taken to Goroka, where he was given pills, though he did not know the nature of the pills. He had been sane since his discharge, but was concerned when we talked that he might be liable to another inexplicable outburst at some future time. He did not mention the alleged rape, but a bible school staff member told me privately that the rape was performed 'in the open in front of witnesses'.

The case illustrates the most common pattern of psychiatric incarceration in PNG. It begins with police intervention after violence and continues with immediate medication at the psychiatric ward. No attempt was made to investigate 'cultural' factors in this short-lived and uncharacteristic outburst. While in the event the young man undoubtedly needed to be restrained, it is questionable whether clinical psychiatric services *per se* were needed. The psychiatric ward appears in the circumstances to have been used as a simple confinement device.

A more florid case in the Goroka Hospital files was the admission of a patient from the upper Nebilyer valley (a valley alongside the Kaugel; the upper part was served by the Tambul administration post). The patient was first admitted in the early 1970s after 'running amok' in his village. File notes classified him as schizophrenic and stated that he 'settled in a few days', being given 100 mg of chlorpromazine three times daily. He was discharged after six weeks. Two years later he was readmitted. File notes (fuller than previously) stated that he had destroyed the local medical aid post with an axe and threatened others in his village. He had been overpowered and taken to Mt Hagen Hospital where medical staff had difficulty controlling him. He had been given chlorpromazine, 50 mg hourly, but this 'only partly controlled him'. A dose increase to 500 mg every six hours (!) failed to control him and after final unsuccessful attempts with 400 mg of chlorpromazine three-hourly (a massive dosage), he had been sent on to Goroka Hospital.

At Goroka he was medicated on a régime of 400 mg of chlorpromazine orally four times a day, supplemented by 100 mg injections and 10 ml of paraldehyde. He settled well, but after two days refused his medicine. After four days his 'brother', who had accompanied him to Goroka, said the patient was alright and wanted to take him home. The doctor pointed out that the patient occasionally broke into 'singsing' behaviour (i.e. chanting and beating an imaginary drum), but the brother considered this to be insignificant. The patient was not allowed home and his chlorpromazine dosage was increased to 450 mg four times a day. After five days he was discovered to have been throwing his medication away, saying it nauseated him. He and his brother were now constantly asking that he be allowed home. After nine days at Goroka the patient was 'co-operative' and a course of electro-convulsive therapy was started. A file summary stated that he was given six treatments, with an interval of three or four days between each treatment. A chart detailing the shock treatment procedure recorded that the voltage was set at 120 volts for 0.9 seconds each time (see below). The first and second treatments involved three attempts each to achieve a convulsion. The fourth treatment achieved no convulsion after three occasions. Thus the recorded six treatments involved a total of twelve shocks. The file did not indicate whether the ECT was modified (i.e. whether the patient was sedated before being given the shock) or unmodified. The man was sent home two days after the final ECT treatment.[7]

The case illustrates a number of factors contradicting the ideal of culturally sensitive psychiatry. From the beginning a preoccupation with controlling the patient's behaviour overrode all 'cultural' considerations. Part-way through his hospital stay the patient's kinsman indicated that the 'singsing' behaviour was not particularly cause for concern and wanted to take the patient home; this 'cultural' assessment was overruled. Finally, when he was calm, the patient was unaccountably subjected to a course of ECT. On this last aspect of treatment, the ECT procedural record is revealing. Orthodox practice in the administration of ECT at that time followed a systematic pattern, if done 'by the book'. Ideally, the voltage was set initially at 100 volts and the timing at 0.5 seconds. If a convulsion was not induced, normal procedure was to increase the voltage by 20-volt stages up to 140 volts. If this still did not induce a convulsion, the time was increased in 0.2-second stages up to 0.9 seconds. Setting the initial voltage at 120 volts and the timing at 0.9 seconds in this patient's case indicates that in the interests of ensuring an immediate convulsion the cautious gradation of voltage and duration was abandoned. Far from having his cultural milieu taken into account, the patient appears to have been reduced at Goroka to an object of psychiatric practice.

I was unable to find a man with the name recorded in the Goroka file in the Tambul subdistrict (the upper Nebilyer was beyond the ethnographic area in which I usually operated, which increased the problems of trying to locate an ex-psychiatric patient twelve years later informed, probably, with an incorrect name).

A third case from the early 1970s illustrates a practical incongruity of the investigation of 'cultural' factors in psychiatry. A woman judged to be about forty years of age was taken into custody after attacking people in her 'settlement' in the Kaugel Valley. She was taken to Mt Hagen Hospital and a European welfare officer investigated the background of the affair. His report was informed by testimony from the woman's husband, his two other wives, four other women, three other men and the European priest at the nearby Catholic mission. Her behaviour had apparently been odd for some years, and she had once been locked up for fighting with another woman. The community considered her mad, she chased people, and she had speared her husband with a stick. She had acted strangely after the birth of each of her four children, on one occasion threatening two European mission workers who took her baby away after she was reported to have threatened to kill it. She fought with her husband's two other wives. The community wanted to 'get rid of her' and the mission backed them. The woman commonly fed the mission's pigs, claimed they were her own and tried to prevent the mission from selling them, physically on occasions.

The welfare officer then interviewed the woman, who was reported to be well-behaved in custody. She claimed her husband and the community teased her and called her 'longlong', and that her husband had engineered her present internment. The welfare officer finally opined that she might be mentally retarded and that her aggression toward villagers was 'a defence mechanism against insults and jeers'. He recommended that she be sent to the Goroka psychiatric ward. She was admitted, and stayed there under a regime of 100 mg of largactil four times a day and 10 mg of valium four times a day. During the five-and-a-half-week stay she was a model patient apart from one altercation with other female patients.[8]

In this well-documented file entry the officer's investigation revealed some violent confrontations and a suggestion of post-natal depression but the driving factor in the woman's incarceration was clearly the woman's social disruptiveness. The community and the local mission wanted rid of her. The welfare officer's assessment of mental retardation and retaliation against teasing in this context suggest that sending her to Goroka may have been a strategy to ease the social tension rather than a specific psychiatric referral. While the welfare officer's investigation could be superficially represented as taking 'cultural' factors into account, it amounted in fact to a review of the social events and an assessment of what was to be done to defuse the situation.

The first or 'Christian' name given for the woman in the hospital file was one that was not uncommon among women of the older generation (in 1986) in the Kaugel Valley. The recorded 'surname' was a puzzle to local people when I asked in the area of the Catholic mission. General opinion was that the woman I sought must have died. As the woman in the hospital file had shared a husband with two others, suggesting that the husband may have been reasonably prominent (with access to resources to 'finance' three marriages), I found the community's lack of recollection intriguing, but I could find nothing further about this case.

A fourth case in the file concerned a woman who spent two months in the Goroka psychiatric ward in the mid 1970s, during which time the hospital had no information on her background or why she was there. She was referred to by five different names. Her age was given as about twenty-five. During her stay in the ward she was medicated with chlorpromazine (75 mg three times daily). Notes described her as quiet, independent, withdrawn and often giggling. An observation on one day described her as acting in a 'sexually provocative' manner, but gave no details. She was still a mystery to hospital staff when she was discharged.

The file contained a letter dated the day after her discharge, giving a background retrospectively to the case. The letter carried no clue to its origin. It identified the woman as a mother of two from a village in the Tambul district. She had been detained by police on charges of prostitution in Mt Hagen. According to the letter, her husband (a labourer working in the Mt Hagen area) had left her because of her irrational behaviour and promiscuity, which had developed after the birth of her second child. The husband had taken the children with him and she had gone back to her parents in the village. She had spent much of her time loitering around the Tambul station and consorting with men, and had eventually drifted into Mt Hagen, where she was apprehended by the police. She was described as having no schooling. The letter stated that she had been admitted as a psychiatric patient because 'her behaviours were odd, like being angry towards any-one without reasons, would cry and would fight other people without reason'.[9]

I encountered this woman a number of times during my fieldwork, and she will be discussed (under the fictional name 'Wanpis') in chapter 4. A vagrant, she survived by stealing food from gardens and eating scraps tossed to her around Tambul station, where she was often to be seen. Withdrawn, uncommunicative and occasionally giggling with no apparent cause, she was virtually inaccessible for conversation. She was by all accounts a sexual target for men – who possibly would have paid her in food[10] – and as a result had given birth, it was said, to several children. Most of these had reportedly died through neglect, though at least one was saved and nurtured by another woman (her husband had left her, as the file letter claimed; he had left the district altogether). Whatever the real historical circumstances of her social fate might have been, the psychiatric encounter in this woman's case had been completely devoid of 'cultural' considerations and had certainly done her no good at all.

By the mid 1980s positions in the medical and law enforcement institutions had largely been localized (exceptions being very senior positions). The chain of medical officers, welfare workers and police staff between Tambul and the psychiatric ward at Goroka Hospital was now Papua New Guinean. The one exception to this during my ethnographic period was the English specialist medical officer who was temporarily overseeing the Goroka psychiatric ward. In terms of psychiatric diagnosis and treatment, the localization of positions previously held by Europeans had not provided the cultural sensitivity that the literature of transcultural psychiatry implies, as some cases from the 1980s show.

The first of these occurred shortly before I arrived in the Kaugel Valley and involved a man who had long been regarded as mad by the local populace because of his strange speech and behaviour, and his occasional outbursts of violence. The particular sequence of events that led to his admission to the Goroka psychiatric ward began when his wife, who was fearful of him, appealed to the medical officer at Tambul health centre to do something about him. She was trying to annul her marriage to him but he refused to co-operate. Village court officials would not help her, she claimed, because they were afraid of her husband, a large strong man. The village court officials, on the other hand, said they were unable to intervene because the custom marriage had been followed by a Christian marriage through the Catholic Church, which put the question of a divorce beyond their level of jurisdiction.

Eventually the medical officer appealed to the Assistant Provincial Community Development Officer, saying that something needed to be done about the man but that the Tambul medical authorities were unable to intervene themselves. The medical officer claimed it was impossible to give the man a tranquillizing injection because he seemed too dangerous.[11] The community development officer spoke to the wife, after which he wrote to the medical superintendent of Mt Hagen Hospital saying the man had a wife and children who always lived in fear of him because of his unusual behaviour, which involved chasing people with an axe, cutting down trees, killing village pigs and generally making a spectacle of himself. This had been going on for the past five years, according to the officer. Following this communication, the head psychiatric nurse from Mt Hagen Hospital visited the Kaugel Valley, spoke to the man, and decided after a thirty-minute interview that he was 'seriously mentally ill'. He spoke to others in the community and learned that the crazy behaviour had developed after the man returned from employment at the copper mine in Bougainville. He had previously been sociable but now isolated himself. There had been complaints about damage and destruction, killing pigs, chopping down trees, destroying gardens. At the time of the interview, wrote the psychiatric nurse in a report, the man was 'trying to cover up his mistakes and says there is nothing wrong with him. Whatever people in village tell is false. He seems to know everything and he is clever. He makes simple words into hard words. Saying that Americans are professors. Some, they come and talk to him'.

No action was taken, but a report from the psychiatric nurse noted that a brother of the alleged madman's wife claimed that he was getting worse. Finally, a few months later after he had refused to go 'voluntarily' to Mt Hagen Hospital, he was taken there by a contingent of police, through arrangement with the ADC. He was not kept at Mt Hagen but quickly transferred to Goroka. He was not violent or aggressive at the psychiatric ward, and was put in a normal patient's room (i.e. with a locked door, a bed and a window with a security grille). It was planned to start a régime of chlorpromazine three times a day, but the day after his admission the patient tore the grille from the window of his room and escaped, making his way eventually back to the Kaugel Valley. No attempt was made to apprehend him.

I had considerable interaction with this man during my stay in the Kaugel Valley, and I saw none of the violent behaviour suggested by the reports given above. He will be discussed at length (using the fictional name 'Hari') in chapter 6, where some of the above testimony will be shown to be misleading and in need of qualification. In the context of 'transcultural psychiatry', we see from the evidence presented that his brief stay in the psychiatric ward owed less to concern with his 'mental health' than to his wife's efforts to be rid of him one way or another. My enquiries into the episode revealed that his wife was not living with him at the time she complained to the doctor, and she had taken their two children with her. She was not, in fact, in any physical danger, but there was friction over custody of the children (the man was attempting to take legal action). When the psychiatric absconder arrived back in the valley no attempt was made by the local community or authorities to have him sent back to Goroka. The ADC at Tambul was replaced at about that time, and the new ADC was a local man related to the alleged madman's wife. Concerned not to show clan bias in local affairs he referred the marriage issue to Catholic authorities in Mt Hagen and channelled the custody conflict into the formal legal system, thereby putting himself in a neutral position and pleading that things were out of his hands. He told me privately that as long as the 'mental patient' did no harm there was no need to interfere with his freedom. Like the third case above, the psychiatric referral of this man was a matter of social tensions and expedience, showing the 'transcultural' aspect of psychiatry again to be an incongruity.

A case in which 'cultural' factors abounded was that of a business college graduate who returned home to the Kaugel Valley and immediately became embroiled in a family conflict that led to his admission to the Goroka psychiatric ward. He came to the attention of the authorities when he was taken to Mt Hagen Hospital by kinsfolk. He was aggressive and refused all treatment. A report that accompanied him to Goroka stated that he was 'talking inappropriately, behaving in a strange way and completely disoriented for place, person, name and so on'.[12] He was given chlorpromazine at Mt Hagen – 200 mg four times a day – but after 1,200 mg given orally and intramuscular doses totalling 1,000 mg he refused further medication. The Mt Hagen report said he had returned from business college two or three weeks previously. His relatives suspected he may have caught malaria (the inference is dysfunction from cerebral malaria) but a medical examination disclosed no traces of the sickness.

He was transferred to Goroka, where he was seen by the specialist medical officer (European, and not a psychiatrist) in charge of the psychiatric unit. The doctor found 'no evidence of hallucination or disordered thought'. He learned from the patient that there had been a conflict with his family over marriage. His parents had chosen a bride for him but he was unwilling to marry her. The doctor referred the patient to the Goroka-based Christian Institute of Counselling (CIC) (see chapter 1, this volume) that employed Papua New Guinean counsellors.[13] The encounter was not entirely successful; the patient was tired of questioning and became irritable. The counsellor concluded that he was not mentally sick, but angry with his family. The patient was returned to the psychiatric ward and maintained on a régime of chlorpromazine, 100 mg morning and afternoon and 200 mg at night. He had been quiet from his admission until the CIC counselling session. Two days after his admission he was reported to be hallucinating and unco-operative, and tried to abscond. He then became quiet again and for the next week was said to be improving, though he began objecting to being kept in the ward and talked of going back to the Western Highlands. He finally absconded, after eleven days at Goroka Hospital. The doctor noted in a final file entry, 'I think he is still sick and definitely has disordered thinking, with paranoid delusions'.

I learned of this episode of incarceration shortly after it had occurred and went to visit the man after his return to the Kaugel Valley. I found him working in a garden with other members of his family and showing no sign of distress or disorder. He confirmed that there had been conflict over a marriage that his family had tried to arrange, but things had calmed down now. After two years at the business college, his outlook had become very different to that of his rural family. He had no interest for the moment in marriage, and certainly not to the woman his parents proposed. There was a mutual lack of understanding between himself and his parents that he found frustrating. The marriage conflict was not the only factor, however. He had tossed some old 'useless' schoolbooks onto a fire in the presence of his family. His family valued formal education highly, though having none themselves did not understand fully what was involved, and became alarmed by his book burning. The incident compounded the tension over the marriage arrangement; his family began saying he was crazy (our conversation was in English, but he used the popular Tokpisin term for madness, 'longlong'). They were badgering him, which increased his anger and frustration. 'They were trying to arrange things for me, and I wanted to make my own decisions', he told me. 'No one asked me how I felt.'

His family had converted to the fundamentalist teachings of the nearby bible school (mentioned in the first case above), and became obsessed with a belief that Satan had taken possession of him. At one point he was made to attend an exorcism ceremony for himself at the bible school. His frustration and anger built up in the face of all the misunderstanding. 'In the end I started saying stupid things and I was violent, so my brothers decided to take me to hospital.' He retrospectively blamed 'traditional ways' (i.e. his parents' attitudes in conflict with his own), rumours and misinterpretation for his demise: He had told everyone to leave him alone early in the episode, but they would not.

In review, the processes involved in this man's psychiatric admission followed a by-now familiar pattern. The priority of Mt Hagen Hospital staff was to control his behaviour; he was not accessible for an assessment of his subjective perceptions. It is likely that the kinsfolk who brought him struggling to the hospital could only offer the information that he had returned from elsewhere and then became 'longlong'. It was not until he arrived at Goroka that an attempt was made to investigate the background; he was quiet by this stage and the doctor was able to ascertain that there had been some kind of conflict with his family over marriage. The doctor took what was ostensibly a constructive step in involving the CIC, but the questioning only served to regenerate the patient's anger. Ultimately, the CIC failed to find a fuller explanation than what was already known, though the counsellor did suggest that the patient was angry rather than mentally ill. This assessment did not affect his treatment; he was simply given chlorpromazine until he absconded.

The Goroka episode may at least have provided a chance for the man (and his family) to cool off, for there was no discernible conflict when I spoke to him, and his family told me he was alright again. As we have seen, there were a number of factors involved which could be glossed as 'cultural' – the young man's battle with his parents' traditional attitudes toward marriage (interestingly a reversal of the 'stress of civilization' scenario), the complicating factors of his family's concern over his burning what they saw as the symbols of formal education, the devil-possession belief[14] – but the institutional processes precluded the possibility that they would be taken into account.

A final example shows again that control of a patient's behaviour from the outset is the overriding preoccupation in the psychiatric encounter, often to the exclusion of constructive assessment of the patient's problem. Early in my field-work the medical officer at Tambul drew my attention to his aid post orderly[15] at Alkena, a place in the valley about 15 kilometres from Tambul station. The orderly, he said, suffered periodic 'mental' breakdowns and had applied for leave to voluntarily attend the psychiatric ward at Mt Hagen Hospital. Shortly after-wards I went to the psychiatric ward during a trip to Mt Hagen and spoke to the patient, an early middle-aged man whom I will call Gawa. According to ward or-derlies he was being medicated with 75 mg of amitriptyline (an anti-depressant) twice daily. This was a fairly high dosage, the usual maintenance dosage being 50 to 100 mg daily.

He told me his problem was not chronic, but he suspected it was related to anxiety. The 'attacks' followed a pattern: They began with a headache that in-creased in intensity, double vision developed after a while, he was unable to sleep and then became crazy. He had been active in community work, had been a com-munity council member, a church worker and had considered trying to move into provincial politics. The 'breakdowns' had resulted in his scaling down his activities until he was an aid post orderly. I was unable to understand why, as a voluntary patient, he was being maintained on such a high anti-depressant dos-age (he was quite groggy from it), but the ward orderlies were unable to enlighten me, and the officer in charge of the ward was not in attendance.[16]

A few weeks later I visited Gawa at Alkena. He had been given amitriptyline to take when he left the hospital and was assiduously keeping to the prescribed dos-age of 25 mg once a day. Since the attacks were only occasional, I wondered at the time what the point was of keeping him constantly medicated, rather than simply giving him something he could take whenever symptoms occurred. I had no en-counters with Gawa for a few months, then saw him one day looking extremely sluggish and heavily drugged. He had recently been discharged from the hospital at Mendi (about 42 kilometres southwest of Tambul) after having a 'breakdown' while visiting kin in that town. He was subsequently taking prescribed medica-tion of chlorpromazine (100 mg morning and night) and amitriptyline (25 mg three times a day). He said he was having difficulty doing anything under the influence of the medication – even working in his garden was an effort.

Concerned by the extreme medication I made a trip to Mendi, instead of Mt Hagen, to buy some fieldwork supplies and visited its hospital to find out what had happened to Gawa there. One of his classificatory sisters (father's brother's daughter) was working as a trainee nurse at Mendi Hospital and had been involved in his admission. She said he had developed a headache while staying with a relative (a Southern Highlands Province politician). The headache had persisted for a couple of days, then his speech had become disordered, he had become angry and violent, so he was brought to the hospital. She told me his 'breakdowns' had begun eight or nine years previously and she confirmed that they were not chronic. The attacks were months apart, and one interval had been as long as two years.

Gawa had been admitted to Mendi Hospital via the outpatients department as a 'violent' patient, and was given 400 mg of chlorpromazine. A (European) doctor told me, 'When they're like that the first thing we do is knock them right down with chlorpromazine'. Hospital doctors had been told by Gawa's sister that he had previously been in the psychiatric ward at Mt Hagen Hospital. On the basis of that information and Gawa's disposition on admission, a doctor (who was not a psychiatrist) had applied a working diagnosis of schizophrenia and had kept him in a general ward (the hospital had no psychiatric ward) under observation, medicated on 100 mg of chlorpromazine and 75 mg of amitriptyline twice daily, for six days before discharging him. The doctor became defensive about the schizophrenia diagnosis (which had gone on record) under my questioning and eventually conceded that this may have been wrong. I proposed that as the maintenance dosages of anti-psychotic medication that Gawa was subsequently taking were debilitating and probably inappropriate they should be discontinued and an alternative, more symptom-specific, medication substituted. The doctor agreed. I obtained from the hospital a supply of an analgesic with a mild sedative effect, of a kind commonly taken in Western societies by people suffering migraine-type symptoms, and returned to the Kaugel Valley.

I explained to Gawa that he had been over-medicated and possibly misdiagnosed, then weaned him off the chlorpromazine and amitriptyline (warning him to expect a sleepless night or two as the tranquillizing drugs left his system). I gave him the new tablets, telling him to simply take two and lie down if he developed a bad headache. My rationale was that since the symptoms were serial and always began with the headache the painkiller combined with a short period (2 to 3 hours) of drowsiness might be enough to offset the development of the later symptoms in the pattern, and the medication need only be taken in the event of an 'attack'. I had no idea, of course, when Gawa might have another attack, but as it happened my theory was tested shortly before I left the valley at the end of my fieldwork. Gawa sought me out (Alkena was on the opposite side of the valley from my host clan's territory) to tell me the good news. He had developed a bad headache of the kind he had suffered previously, took the pills, had a short nap and felt better. He was pleased with the new medication and surprised at the simplicity and effectiveness of it (as much as the single application can be taken as a real test), after his experiences with major tranquillizers and anti-depressants.

The 'schizophrenia' diagnosis (by a non-psychiatrist) and its pharmaceutical aftermath certainly appear to have been unnecessary and excessive, but it must be said that the investigation of 'cultural' factors in Gawa's case probably would not have been particularly enlightening. The episode is recounted here to demonstrate that in practice, as with other cases cited above, events themselves often reduced psychiatry's role to that of social control so drastically that no factors beyond the patient's immediate manageability, let alone 'cultural' ones, were taken into account.

Culture in Theory and Practice

At the beginning of this chapter I briefly referred to changing conceptions and definitions of culture in anthropology and in psychiatry. In anthropology these have come to a point where some contemporary anthropologists have even advocated that the concept be discarded on the grounds that it is analytically inadequate (for example Kuper 1999). Accordingly, references to 'culture' in this chapter have often used quotation marks, intentionally to disturb the continuity between that concept and the rest of the text. In the case of psychiatric discussions of culture in PNG during the period covered here, there are certainly serious conceptual inadequacies. In one instance Burton-Bradley conflated 'culture' with 'custom' (Burton-Bradley 1973a: 4), an equally ambiguous term; but more generally 'culture' remained amorphous in his work. Wittkower and Prince (1976) managed little better: 'Culture may be defined as the whole fabric of ways of living that distinguishes one human society from another. It is the blueprint for living that presents individuals of a society with modes of behaviour, thought and feeling' (1976: 6). Other texts produced in the 1980s tended to approach the 'cross-cultural' context instrumentally: 'The most basic question a mental health service provider must ask in a culturally different setting is "When do you adapt the assessment diagnosis and therapy methods of a dominant culture and when do you substitute the unique approach of an indigenous culture?"' (Pedersen, Sartorius and Marsella. 1981: 11). Thus culture was predetermined, by the nature of the exercise, as whatever made a group of people different. At the beginning of this chapter I cited the recent problematizing of the culture concept in transcultural psychiatric discussion, yet in transcultural psychiatric discourse in PNG 'culture' has continued to connote at best a loose idea of non-European customs and beliefs, rendered more complex by virtue of ethnographically compelling themes like the country's 700-or-so languages and 'cultural diversity'. In terms of this definition-by-inference, the task of the transcultural psychiatrist is unrealistic, and the literature of the late colonial period and beyond has in fact described a practice that remains an ideal only:

> [The psychiatrist] must separate psychiatric symptoms from cultural norm. If he does not know the cultural norm he will need to seek this information from anthropological studies, patrol reports, the kinsmen of the patient, mental health records, and where appropriate from the patient himself. The information from these sources is of variable quality, and the goal of the examiner is to achieve the maximum degree of validity of data that is in fact possible. (Burton-Bradley 1973a: 5)

Since its discursive introduction in the 1950s transcultural psychiatry has undergone a number of developments, such that according to Jatinder Bains (2005) it currently means different things to different people. Bains offers four contemporary understandings of 'transcultural' psychiatry. Firstly, it is an attempt to apply modern Western concepts of disease to non-Western peoples; secondly, it is an attempt to understand illness in terms of 'local cultures'; thirdly, it is a form of psychiatry related to the concept of racism and advocating the rights of ethnic minorities; and fourthly, it is a form of practice that 'in some senses fuses all of the above' (2005: 151). Yet Bains' four kinds of transcultural psychiatry – and significantly, for our purposes, his second 'meaning' – leave unaddressed the processual flaw which is implicit in the case examples I have given above. The theoretical models discursively situate the transcultural psychiatrist at the interface of the medical institution and the society of the patient. From this medial position the modalities referred to earlier in this chapter – cultural relativism, distinctions between organic and non-organic psychoses and between real psychosis and psychodrama – seem to be natural extensions of psychiatric practice. However, as we have seen in the cases cited in this chapter, the psychiatrist does not in fact occupy this privileged analytic place, and is not in a position to seriously take 'cultural' factors into account. Culturally sensitive psychiatry remains a paper concept.

The cases reveal that the psychiatric encounter usually follows socially disruptive behaviour and becomes part of a sequence of attempts to control that behaviour. The psychiatric treatment (mostly preoccupied with drugging and retaining the patient) proceeds under its own institutional impetus, regardless of 'cultural' factors which may be presented to it. Indeed in one case above we saw that a man was detained, and even given shock treatment, after a kinsman had shrugged off behaviour perceived by a non-Melanesian as odd and had asked that the patient be released; another was detained after a Melanesian counsellor had suggested that he was not 'mentally ill'.

The examples from the mid 1980s indicate that psychiatric diagnostic and treatment practice had not been affected by post-colonial changes in policy or personnel. Law enforcement officers invariably mediate between the community and the psychiatric institution (since violence is usually the precipitating factor in admission) and the fact that these officers were now always Melanesian did not significantly alter their role as the agents who removed the socially unmanageable and violent person to the hospital. Contract hiring of expatriate personnel and constant shuffling of staff from one hospital to another under the exigencies of volatile administration politics, on-the-job training schemes and staff shortages resulted in minor variations in practice within the psychiatric unit. For instance, the doctor who had been in charge of the Goroka unit for about eighteen months when I visited the ward during fieldwork had not used ECT during his tenure and followed a European trend of the time in avoiding diagnostic labels (though his junior staff tended to use them indiscriminately). He tried to apply a firm policy that all patients should be accompanied by a kinsperson (who slept elsewhere in the hospital) to aid communication and sympathetic management of the patient. The latter policy did not succeed in avoiding the 'dumping' of individuals into the ward by authorities, however. Further, his relatively liberal policies such as the use of counselling services did not prevent the ward from acting as the final repository of people who, as we have seen here, could arguably have been treated differently.

In more recent times there have been no indications that the practice of transcultural psychiatry (however one conceives it) has become more competent, in terms of the incorporation of 'cultural' factors. A study in 2001 found that Papua New Guinean health workers operating from hospitals and rural health centres, with minimal training in mental health, had little confidence in 'culture specific' diagnosis of 'mental' disorders, but were more comfortable with 'modern' (i.e., quasi-psychiatric) diagnosis (Koka, Deane and Lambert 2004). The study itself was conceptually informed in part by the colonial-period writings of Burton-Bradley (and even Sinclair) and implicitly took culture, ahistorically, to be a locally bounded entity typified by a specific set of practices (2004). Psychiatric diagnostic categories in PNG are informed by standard international classifications of 'mental disorders' such as the American Psychiatric Association's *Diagnostic and Statistical Manual of Mental Disorders* ('DSM') series (APA 1994). The most common diagnosis applied by professional psychiatrists in PNG, from Burton-Bradley's heyday to the present, has been 'schizophrenia' (see, for example, Burton-Bradley 1969; Johnson 1997). An optimistic view of the relationship between anthropology and psychiatry in the twenty-first century has been offered by Allan Young (2008), who suggests we can move beyond a number of critical views of psychiatry that were prevalent in the 1970s. These included the idea that psychiatric institutions and practices promoted hegemonic cultural perspectives and social interests (ibid., 299). I have found little evidence to suggest that Young's optimism can yet be extended to the consideration of psychiatry in PNG.

The contradiction between psychiatric theory and practice in PNG has been elaborated in this chapter through some brief case studies that show that the social control function of psychiatry predominates over its theoretical presentation as a mental health service. This function prevails in the face of developments in the theoretical discourse that emphasized a 'transcultural' dimension and implied the possibility of psychiatry being applied with cultural sensitivity. While the discourse positioned psychiatry as face-to-face with the society in which the madness occurred, this has not been the practice. Further, individuals arrive at the psychiatric unit not primarily because they are mad, but because of particular precipitating factors such as violence or social disruption. Thus madness, in its social context, remains largely beyond the practical gaze of psychiatry.

To understand further the relationship between psychiatry and the phenomena it is attempting to address requires a shift in focus to an ethnographic context, which is the intent of chapter 3.

Notes

1. For a range of perspectives on the non-psychiatric influences on diagnosis and treatment, see Banton et al. (1985), Basaglia (1980), Brown (1985), R. Castel, F. Castel and Lovell (1982), Clare (1976), Deleuze and Guattari (1977), Laing (1973), Laing and Esterton (1970), Szasz (1972, 1973), Turkle (1978), and Scheper-Hughes and Lovell (1987).

2. The enduring journal *Transcultural Psychiatry* began publication in the 1960s, and for a sample of publications evidencing increasing acknowledgement of 'cultural' factors from that period, see Burton-Bradley (1965a, 1973a), Langness (1965), Lebra (1976), Leff (1981), Pedersen, Sartorius and Marsella (1984), Simons and Hughes (1985), Westermeyer (1976), and Okpaku (1998).

3. The anecdotal substance of functionalist and quasi-functionalist ethnographers often attested to the chronic tensions in their subject groups (see, for example, Fortune 1963; Bateson 1972: 97–106), but was negated by the formal imposition of an institutional structure (kinship, religion, economics, etc.) which fragmented the human interaction in the interests of a workable descriptive model.

4. Burton-Bradley published two articles in *The Medical Journal of Australia* in the early 1970s psychiatrizing 'cargo cults' and their leaders (Burton-Bradley 1970, 1973b).

5. These are constituents of the 'village court' system mentioned in chapter 1.

6. Goroka Hospital Patient File U128.

7. Goroka Hospital Patient File U302.

8. Goroka Hospital Patient File U161.

9. Goroka Hospital Patient No. AD114-723. No file number.

10. Two Kina (Aus $4 in 1986) was the commonly spoken-of payment for an act of non-professional prostitution in rural PNG at the time – hence the Tokpisin phrase 'Tu Kina bus meri' (K2 in-the-forest woman) – though more was frequently offered or charged. In respect of this particular woman the act could hardly be categorized seriously as prostitution, though people did refer to her as 'Tu Kina bus meri' or in the local language (Umbu Ungu) as 'wapera', meaning a promiscuous woman.

11. Correspondence quoted in this case was filed at Goroka Hospital, file unnumbered. The man was patient AD194-988. Details in this case are partly from the file and partly from conversations and research in the Kaugel Valley.

12. Goroka Hospital Files, patient number AD 202-084; file unnumbered.

13. 'Barefoot counsellors' was a CIC catchphrase.

14. The highlands of PNG were flooded at the time with both orthodox and exotic missions. It was not unusual to find people in rural communities who had developed mixtures of 'incompatible' Christian beliefs such as conventional Catholicism and fundamentalism; devil-possession was a common cry.

15. An aid post consisted of a single-room hut stocked with aspirin, penicillin, anti-malaria pills, etc. – a first-aid resource, in other words. The orderly usually lived nearby, was trained in taking temperatures, giving injections, etc., and could deal with minor health problems.

16. I was never able to locate this person despite several attempts during my fieldwork period. He was never at the hospital when I visited.

3

MADNESS AND THE AMBIVALENT USE
OF PSYCHIATRY IN THE KAUGEL VALLEY

A diagnostically cautious psychiatrist would have found very few cases of serious mental illness in the upper Kaugel Valley in the mid 1980s. In the duration of my fieldwork I knew of only two or three people who I considered might be classified by a psychiatrist as having a psychotic condition of the order of, say, schizophrenia or manic depression. A handful of others might have been diagnosed as having minor, or temporary, psychiatric disorders, but the number was negligible among the population of more than 20,000 at the time. The foregoing declarations, of course, need a great deal of qualification, not least because 'traditionally' the people of the Kaugel Valley – the Kakoli – did not subscribe to the conception of madness represented by psychiatry. In this chapter I elaborate the history and sociality of the Kakoli in some detail, and discuss their conception of madness and their attitude toward psychiatry.

The Kaugel Valley before European Contact

The upper Kaugel Valley is on the border between the Western Highlands Province and the Southern Highlands Province. It is about 20 kilometres long and about 9 kilometres across at its widest points, and runs from north to south. Its western wall is provided by Mt Giluwe, the second highest mountain in PNG, an unactive volcano that rises to almost 4,600 metres. During the day the upper part of Giluwe is often rendered invisible to the valley dwellers by cloud and mist. The Kaugel River meanders along the valley's floor, about 2,200 metres above sea level, gathering momentum for a descent into the Southern Highlands Province.

The term 'Kaugel' is an Anglophone approximation of the name of the river and the valley people. They traditionally call themselves *Kakoli* (identifying themselves geographically with the river), and are referred to as such by their neighbours to the southwest, the Mendi. The people of the neighbouring upper Nebilyer Valley to the northeast refer to them as the *Kakuyl* (Merlan and Rumsey 1991: 22). The Melpa of the Mt Hagen area to the west call them the *Gawigl* (Blowers 1970: 1), and the Kumdi Engamoi to the north call them *Gawul* (Brandewie 1981: 32). Other recorded terms used to refer to the valley have included *Kâwudl* and *Kakudl* (Strauss 1990: 6 and *passim*). European explorers of the highlands in the 1930s referred to the river and the people variously as *Gauil* (Leahy 1936: 250) or *Kagole* (Champion 1940: 192) or Kaugel. I will refer to the valley as the Kaugel and the people as the Kakoli and I will generally follow the orthographic convention of their first ethnographer (Bowers 1968), which differed slightly from that of her roughly contemporaneous linguists (Blowers and Blowers 1969; Blowers 1970), in my linguistic representation (see appendix A).

The language of the Kakoli is *Umbu Ungu*.[1] It is a dialect in the chain of languages that Wurm (who referred to it as 'Gawigl') called the 'Hagen sub-family' of the 'Hagen-Waghi-Jimi-Chimbu' family of the 'East New Guinea Highlands Stock' (1961: 114–15). Other linguistic classificatory methods have been used in the region. Brandewie, for example, includes the Kakoli ('Gawul') with the Melpa, some Wahgi groups to the east, and some Jimi groups to the northeast in a linguistically similar group collectively termed *Mbowamb*[2] (Brandewie 1981: 32–33; cf. Strauss 1990). More recent attempts to group the non-Austronesian languages of the PNG highlands are exemplified by 'Trans New Guinea Family', in which languages of the Western Highlands Province are included (see, for example, Foley 2000: 363).

Proceeding southward down the Kaugel Valley from its head, Umbu Ungu has three sub-dialectical variants: one used by clans at the valley's head, another used by clans in the main part of the valley (where I resided), and a third by clans in the section where the river begins to descend toward the area known as Ialibu (where a related language called *Imbonggu* is spoken) in the Southern Highlands Province. Also, as Bowers noted (1968: 19), a variant of the language of the eastern Enga groups of the nearby Lai River area is used by some clans in the northwest corner of the valley. These linguistic variations can perhaps be considered in relation to some observations by Bowers that in matters of self-identification by the Kakoli, '[c]ultural identifications ... vary with the context' (ibid.), that the Kakoli 'recognize and name neighbouring peoples according to cultural affiliations' (ibid.), and that when Kakoli identified people around them 'geographic considerations are more important than linguistic affiliations in placing peoples culturally' (ibid., 21). Underlying these observations about language, geography and identification are questions about the way the Kakoli came to be living in this high-altitude valley, which is climatically and agriculturally unsympathetic to human habitation.

Nancy Bowers was the first ethnographer of the Kakoli. She specialized in agricultural anthropology, and her unpublished doctoral thesis was a study of the effects of human subsistence activities on the local ecosystem (Bowers 1968).

Agricultural activity was an important consideration in her attempt to gauge not only the effect the Kakoli had on the landscape, but also their movements before European contact and the amount of time they had been in the Kaugel Valley. In part, she was responding to what was at the time a recent proposition that the sweet potato, *Ipomoea Batatas,* had been a post-Columbian introduction which had quickly and significantly affected the agricultural production, subsistence potential and relative population growth in the PNG highlands. The 'ipomoean revolution' (of about 400 years ago) is nowadays a taken-for-granted factor in discussions of the prehistory of agricultural and pig-husbandry developments in the highlands (see, for example, Feil 1987: 26–31). Its importance for Bowers was that the sweet potato probably reached the Kaugel Valley about 200 years before her early 1960s fieldwork, enabling a very small group of people there to establish stable agricultural production and begin to expand into a significant settled population (1968: 222–28). Before the introduction of the sweet potato, the upper Kaugel had been a difficult environment in which to grow subsistence crops. With a concerted effort to maintain drainage and irrigation systems it was possible to grow taro in some bottomland by the river. Edible pandanus ripened periodically, but only a few edible varieties of leafy vegetables were available and no fruit grew. There was limited game at that altitude (2,200 metres and above). Bowers suggested that only a small pre-ipomoean population could have survived through limited agriculture, foraging and hunting in the valley (ibid., 236–45).

Bowers's cautious suggestion, then, was that while a sprinkling of people may have been present previously, a significant population was able to establish itself beginning about 200 years before her fieldwork (now about 250 years ago) using the sweet potato as its subsistence staple. Subsequent to Bowers's findings, an archaeological discovery – the 'Tambul spade' – in the valley in the 1990s was dated to more than 4,000 years ago (Golson 1997). An inference of the discovery was that someone was swamp-draining and irrigating for agricultural purposes at that much earlier time, which fitted with similar findings at Kuk, a long-established archaeological site at a lower altitude near Mt Hagen (see, for example, Golson 1981). However this single discovery did not provide enough evidence to seriously contest Bowers's thesis, which was elaborated with supporting observations on land use, warfare patterns and visible, measurable effects of human activity on local vegetation.[3]

According to their own mythic accounts, the Kakoli migrated into the upper valley both via its northwest side from an area now known as Enga, and up-valley from the Ialibu area to the south. During fieldwork I was told some *temane* (Umbu Ungu: roughly, 'old stories' or myths)[4] containing descriptions of migration paths, and Bowers passed on to me some that she had been given that had similar contents. Some of these stories describing people's arrival in the valley implied that they brought sweet potato with them to plant and found wild pigs already there.[5] Movement and migration before European contact in the area was due to a variety of factors including climate, disease (precipitating flight, see, for example, Karoma 1978: 39), warfare, and agricultural developments. Oral history attests to prehistoric warfare in the region, including various wars within the Kaugel Valley, from which cleavages and antagonisms remain evident (see, for example, Bowers 1968: 77, 136–40, 167–68; Karoma 1978: 43, 46–47). Severe frosts are known to have periodically occurred in the area before European contact (Allen et al. 1989), and Bowers was told of some temporary outmigrations due to crop-devastating frosts within living memory (Bowers 1968: 30, and pers. comm.).

By the mid twentieth century, the Kakoli were living on both sides of the Kaugel Valley, in dispersed homesteads (i.e. there were no 'villages' or hamlets) in about forty paired patrilineal clans, mostly engaged in subsistence agriculture. When Bowers carried out her initial fieldwork in the early 1960s, the population was about 11,000. By the mid 1980s, when I carried out my fieldwork, it was said to be 28,000.[6] On the western side of the valley, where I resided, the territory of each clan rose from the valley floor more or less straight up Mt Giluwe into a forest, the upper part of which was common hunting ground, although there was little left to hunt. Traditionally the Kakoli were involved in trade networks through which they and neighbouring peoples exchanged items such as salt (which came mainly from Enga), tree oil (mainly from Mendi), stone axes (from Melpa and the Nebilyer Valley), and shells (traded up from the coast through the eastern highlands). Much of this trade died out with the coming of European goods, but the other major form of exchange, expressed through pig-debt cycles and involving the periodic mass slaughter of pigs for distribution, has continued. The Kakoli were traditionally involved in chains of pig exchange connected to the Enga system known as *tee* via the Lai River area (see Wiessner and Tumu 1998: 164, 172, 293–324). These large pig distributions among clans were called *makale* by the Kakoli, and within the valley they contained some variants on the *tee* (see Bowers 1986: 96–104; cf. Feil 1984). Indeed, they fell somewhere between the *tee* and the Melpa system known as *moka* (see Strathern 1971; Strauss 1990: 244–70), indicating some historical interconnections between those two systems (see Feil 1987: 263–68). While the altitude and environmental conditions of the Kaugel Valley dwellers were different from those of the majority of highland communities, their social organization was fairly typical of the region. Extended kin groups were the main units of generalized production, consumption and exchange, and engendered a strong sense of social and familial responsibility.

European Contact and Subsequent Changes

It is possible that the Kakoli heard about Europeans before they saw them, for the explorations and material effects of the latter in the region could not have escaped the wide-ranging communication networks of the highlands. Like many other highland peoples the Kakoli described Europeans not as 'white' but as 'red' people (I was often addressed in Umbu Ungu as *kondole Maikol* – red Michael – or simply as *ye kondole* – red man). No stories suggesting that the Kakoli thought Europeans were the ghosts of their own ancestors were heard by Bowers during her early fieldwork,[7] or by myself during my fieldwork. The first Europeans to set foot in the Kaugel Valley appear to have been members of the Leahy brothers' expedition of 1934 (see Leahy 1936: 256–58), possibly followed by members of Ivan Champion's 'Bamu-Purari Patrol' in 1936 (see Champion 1940: 194). The government anthropologist E. W. P. Chinnery may have flown over the valley in the early 1930s (see Chinnery 1934: 116). During my fieldwork in 1985 a man I judged to be aged in his fifties told me that as a very small boy he had fled and hid in the forest on hearing and seeing an aeroplane overhead for the first time. The Kakoli saw few, if any, Europeans for more than a decade after these incidents, though missionaries and colonial officers were already established in Mt Hagen by the late 1930s.

The first official government patrol entered the Kaugel Valley in 1949, followed by missionaries in the early 1950s. Regular patrols began in 1956 and a permanent Administration post was established in 1957 at a site locally called *tambili*, which Europeans called Tambul, at the head of the valley. A Catholic mission was established at Kiripia, about 11 kilometres downvalley from Tambul, with a primary school attached. Lutherans established themselves on the valley's eastern side at Alkena. An 'East and West Indies Mission' (later renamed the 'Evangelical Bible Mission') school was established at Tambul in the same period, and by 1960 there was an Administration primary school. In time the patrol post at Tambul became a subdistrict headquarters and 'Tambul' subsequently became the name of the administrative subdistrict. As a result the Kakoli of the upper part of the valley are commonly known nowadays as 'Tambuls'. Tambul is still an Administration headquarters; the former subdistrict has become the 'Tambul-Nebilyer District'.

The valley's unfavourable agricultural conditions were recognized early by the Administration. A district commissioner commented on receipt of a patrol report in 1958 that '[t]he area does not appear to have much agricultural future'.[8] An agriculture training centre, 'Yangpela Didiman Senta' (Tokpisin: 'centre for young agriculturalists') was established at Alkena in 1972, financed by the Lutheran Church, partly as a response to the perceived agricultural difficulties posed by frosts in the valley. A research-oriented high-altitude agricultural station was also established near Tambul by the Department of Primary Industries which encouraged the Kakoli to try growing European vegetables (cabbage, carrots, potatoes, etc.). While these various enterprises held potential for the Kakoli to develop their agricultural production toward successful cash-cropping, other factors acted as obstacles, and optimistic initiatives (including the introduction of pyrethrum production in the mid 1960s) came to little. One problem was access to major centres. Tambul is about 32 kilometres west of the main provincial centre, Mt Hagen, and a winding road from the town up through the Nebilyer Valley to Tambul was dug in 1959, but was still mostly unsealed by the 1980s. Another road into the valley from the southern end was built in the early 1970s (an enterprise initiated by a local politician), but had deteriorated and was unusable by four-wheeled vehicles by 1985. An airstrip had been cleared at Tambul when the patrol post was set up, but was poorly oriented (according to local memories, small planes were buffeted by crosswinds from both east and west when landing and taking off, and accidents were common) and boggy, and had deteriorated by the 1980s from disuse.

The difficulties of high-altitude subsistence production, heavily reliant on the sweet potato, had been reduced with the introduction of imported tinned fish and rice, which by the 1980s were being sold in small tradestores at Tambul and at points along the road down the valley. But cash to pay for these simple luxuries was scarce due to the relative isolation of the Kakoli from Mt Hagen, the nearest place where a significant wage income could be had. Young men in particular went to work on plantations elsewhere in the highlands or made their way to the coastal towns in search of employment and their remittances, or their return with distributable funds, were an important but unreliable source of general income in the first post-colonial decade. With its old, prefabricated, colonial-era buildings and struggling medical post, Tambul was fairly typical of rural government outposts in the 1980s, and Kakoli of my acquaintance expressed a general sentiment of disappointment at the failure of *divelopmen* (Tokpisin: 'development' – which means, to rural Papuan New Guineans, tangible material benefits) to 'arrive' since the end of the colonial period.

Madness According to the Kakoli: A Hermeneutic Problem

The majority of Kakoli were nominally Christian (Catholic mostly, but also Lutheran, Baptist, and Evangelical) by the 1980s. Their pre-existing ontology had inevitably been affected by missionary and colonial administrative proscription and changing material circumstances. A significant amount of pre-contact ritual activity had disappeared (this was clearly evident through the reminiscences of adults and the elderly). Regular Sunday church-going seemed to me to be a partial compensation for this, offering a weekly opportunity for people to dress up, sing, and listen to stories and declamations from a priest. The Catholic Church services on Sunday mornings at Kiripia were a prelude to a day-long 'market' in an area adjacent to the church grounds, which gave the opportunity not only for buying vegetables and betelnut, but for promenading and gambling (cards, dart-throwing) and later in the afternoon, sports (mostly basketball and soccer). Church-going and church-group activities were also seen, alongside business and various money-making ventures, as a 'road' to *divelopmen,* the elusive material benefits which according to local expectations would be acquired by the successful manipulation of colonially introduced activities.

Among Christianized Kakoli, God had been added to local ontology as a powerful being who could affect their life and fortunes, who would punish them for doing 'bad things' (*ulu keri:* the Umbu Ungu term used by missionaries as a translation of 'sin'), who could perhaps be persuaded through prayer (the Tokpisin *beten* was used for 'pray'), and who enjoyed hearing hymns and songs of praise which reinforced his prestige and acknowledged his power. Despite the use of Umbu Ungu words for the translation of Christian concepts such as 'God' (*Pulu-Yemo:* original root-man) 'sin' (*ulu keri*), 'soul' ('*mini*', see below), and 'Heaven' (*mulu:* sky, or *mulu-kola:* sky-place), Kakoli seemed to view the principles and processes of the introduced God's agency as differing from those of ancestors and a range of invisible beings who already populated their world (I will return to this observation in the concluding chapter of this volume).

The coming of the Christian God had not, then, eclipsed their pre-Christian ontology by any means. Ancestors were immanent, powerful beings resided in the crags at Mt Giluwe's summit, and sorcerers from beyond the valley could be engaged at a distance to malevolent ends within the valley. People's lives, including their health, could be affected through invisible forces for which the general term in Umbu Ungu was *kuru.* At first I interpreted *kuru* (a term common to a number of languages in the western highlands) as 'spirit',[9] but later decided this was an inadequate translation. I still have not come to an analytically adequate understanding of *kuru,* and prefer to leave it untranslated. We will revisit the term later in this chapter.

The Umbu Ungu word used of madness was *kekelepa.*[10] A crude but convenient sense of this term could perhaps be provided by the English term 'crazy', used in its loose, wide, lay sense (insofar as 'crazy' can mean 'mad' but can be situationally replaced by terms such as 'stupid', 'silly', 'irrational', 'ridiculous', and so on). I have translated *kekelepa* in this way in a previous publication (Goddard 1998), simply to dissuade readers from psychiatric interpretations. Here, though, I must be more precise. The word *kekelepa* is related to the word *kelepa.* The latter term could be translated roughly as 'separate oneself from' or 'leave', as in leaving a group or a place. More exactly, though, *kelepa* was used of an individual who left a group or a familiar location in order to 'do something else'. There was an implication that whatever they did might be singular or removed from what was the normal practice of the group.[11] In the case of the word *kekelepa,* the prefix /*ke-*/ was a root expressing diminution.[12] The degree of estrangement implied by *kelepa* was consequently reduced, in that the *kekelepa* person had not physically left, but his or her behaviour was nevertheless estranged from the 'normal' behaviour of the group. In 'normal' we can include, depending on the social context, anger, violence, laughter, crying and so on. To put this another way, '*kekelepa*' referred to behaviour which was socially 'out of place'. Significantly, when alcohol was introduced into the Kaugel Valley toward the end of the colonial period the Kakoli extended the application of the term *kekelepa* to include drunkenness. While *kelepa* usually implied the departure of a single person from a group, the term *kekelepa* could be used of a group of people all behaving crazily, such as a group of drunkards. No qualifying terms were added to '*kekelepa*' to distinguish between permanent or short-term, or chronic or temporary, craziness. The word *kekelepa* was applied widely, not only to people whom a psychiatrist might consider to be seriously mad and to drunks, but also, rhetorically, to naughty children and even to uncooperative pigs.

The concept '*kekelepa*' cannot be subsumed under a psychiatric paradigm. To the contrary, the social behaviour associated in the West with psychiatric categories of madness was included by the Kakoli in the more general category of 'crazy' behaviour referred to by the term *kekelepa*. Madness was understood according to an ontology that did not include 'mental' illness. On the other hand, there is no doubt that *kekelepa* behaviour among the Kakoli could sometimes involve inadequate thought, or confused thinking. Some elaboration of Kakoli ideas of the person is necessary at this point, for references to 'thought' and implicitly the 'mental' aspects of personhood can encourage a subtle and unnecessary psychiatric turn in Western interpretations of Melanesian (and other) conceptions of madness, as we shall see below.

The Kakoli did not divide the person into the same 'mental' and 'physical' (or 'mind' and 'body') categories that are prevalent in Western societies. For instance, the Kakoli concept *numan* superficially lent itself to translation as 'mind' inasmuch as it referred to something invisible that inhered in the person and thought, intended and willed. But it was not located in the particular mass of soft tissue that Westerners call the brain (and for whom the latter term itself is redolent of 'thought'). In fact it was located in the torso, above the stomach in the area encased by the ribcage. Moreover, as a locus of thought, intention and will, it was socially interactive, able to affect and be affected by others and by relations with others. The *numan* had a contingent relation to the *mini*. *Mini* was the word that missionaries used as an Umbu Ungu translation of the Christian 'soul'. However, I am reluctant to translate it as such, because of the connotations of the English word.[13] The *mini* inhered in the soft tissue of the body overall, and its final departure at (physical) death resulted in the decay of the body's soft tissue, leaving only the bones. The *mini* was what one saw in images of a person disconnected from his or her physical body, such as his or her shadow, or reflection. Images of a person in others' dreams were also his or her *mini*. It could jump away from the body when someone was startled (a state referred to as *mini wale*) and left the person stupefied until it returned. When the body died, the *mini* survived it, remaining proximate for a period before joining the ancestors. The *numan*, however, did not survive the death of the body.[14]

The *numan,* having a social aspect, could not be conceptually reduced to 'thought', for thought was only part of what contributed to a properly developed and healthy *numan.* Thought or thinking (*konopu*) in itself was engaged not only with the *numan* but with locations in the body where emotions were variously experienced (for example, anger in the belly, fear in the chest, and so on). Further, there was a connection between the health or condition of the body and that of the *numan.* The brief sketch I have given here has omitted a great deal of detail about the Kakoli 'person', such as the dynamics of the interaction of *numan* with a corporeous range of emotions, with senses such as hearing, sight and smell and with speech, as well as with the *numan* of others. Enough detail has been given, however, for present purposes. The particular aspect of the *numan* that should be noted in this regard is that it was socially contingent as well as integrated with the individual body by virtue of its physical location in the chest. This in itself should make us wary of assuming that the physical/mental categories implicit in psychiatric paradigms of the (individualized) person were shared by the Kakoli.

In addition to considerations of 'personhood', we must acknowledge a hermeneutic problem brought about by Westerners' conditioned tendency to relate madness to the mind and to apply psychiatric concepts by default. The problem is visible, for example, in Reay's ethnography of the Kuma of the western highlands (Reay 1965), during which she briefly discusses hearing and deafness and says (rightly or wrongly) that the term for 'ear' (*kumugl*) is extended to include deafness and other conditions. She writes: 'The Kuma and their neighbours use this term further, to cover any kind of *mental* abnormality from idiocy to *psychosis*' (1965: 18–19, emphasis added). Similarly, Langness (also a highlands ethnographer), in an attempt to reconstruct the social context of an incident of madness (1965) not only uses the term 'psychosis' in the title of his article, but goes on to define it. He borrows a description from the Sinclair report (see chapter 2, this volume) – 'a state of uncontrollable behaviour usually characterized by inappropriate excitement or withdrawal' (Sinclair 1957, cited in Langness 1965: 259) – which is reasonably sociological. Yet, having created the potential to move away from psychiatric conceptualization, Langness immediately shifts back: 'Hysteria has often been cited as a component in non-Western *mental* disorders, and the term hysterical *psychosis* is not without precedent' (1965: 259, emphasis added). The problem is not confined to PNG ethnography: Noricks (1981) for example makes a serious attempt to explicate the range of applications of the term *fakava-levale* by the Niutao islanders of western Polynesia. He is careful to use the loose translation 'crazy' (1981: 19), yet his article is subtitled 'A Polynesian Theory of Mental Disorder' and he comments at one point that the 'people of Niutao do have a theory of psychopathological disorder ... implicit in their perceptions and categorizations of those around them' (ibid.).

My point is more than semantic. The difficulty for Western anthropologists is grounded in the political history of psychiatry, as a brief review of one or two periods in that history will show. Terms like 'mental illness' and 'mental disability', (let alone the more technical jargon of psychiatry) were not in currency before the nineteenth century in Anglophone countries, nor were their equivalents on the European continent. Pre-psychiatric attitudes to madness in Western history moved through a variety of somatic, theological or demonological explanations and interpretations according to political conditions.[15] With the development of the capitalist mode of production, and the rise of medical science as an economic enterprise,[16] the notion of 'mental illness' gained currency as an interpretation of madness (Scull 1979: 125–63). Modifications in Cartesian dualism provided the metaphysics for the brain to be seen as the medium of non-somatic influences, which legitimated the establishment of a special branch of medical science concerned with 'mental illness' (Cooter 1981; Doerner 1981: 89, 153–56). In England this involved some conflict between 'medical' and 'moral' treatment theories and practices before the institutionalization of mental hospitals with resident medical directors (see, for example, Scull 1979: 158–63; Bynum 1981; Tuke 1964; Mellett 1982).

In the rise of medical science and psychiatry the delegitimation of previous attitudes to, and explanations of, madness was a political contingency of bourgeois scientism. The resulting hegemony of psychiatry, and the jargon generated by the establishment of 'mental illness' as a sphere of discourse, has remained, although it was challenged with the rise of a body of critical literature (sometimes called 'anti-psychiatry') in the 1960s (reasonably summarized, for example, in Bynum, Porter and Shepherd 1985: 1–20). The development of the discourse of 'mental illness' into ahistorical, universalistic usage has been effective to the point where it is even retrojected beyond its actual historical horizon in response to discursive exigencies. A handy example is a historical study of madness and government policy in sixteenth- and seventeenth-century England (Neugebauer 1981), an admirable investigation of social and economic factors involved in juridical practice concerning 'disabled' people of that era. The author's hermeneutic achievements (the exposure of historical worldviews via the textual evidence of the period) are partly undermined in the text, as can be inferred from a footnote inserted for the benefit of contemporary readership: 'Although the term *disability* appears on a few occasions in seventeenth century Wards documents, *mental disability* does not. This phrase will be employed in the present study as a generic term for *lunatics* and *idiots* together. *Mentally subnormal* will be the phrase adopted when referring specifically to *idiots*' (1981: 2n, emphasis in original).

The transjection of 'mental illness' (and derivative 'mental health') language-games into anthropological literature has been effected not only through the dissemination of psychiatric concepts into general usage but also through the psychoanalytic interests of some influential ethnographers (for example, Bateson 1972; Roheim 1950, 1962; Lévi-Strauss 1972; Devereux 1956, 1961, 1980), who drew on the ideas and terminology of Sigmund Freud. While Freud is popularly seen as a psychoanalytic, rather than a 'psychiatric', theorist he developed his ideas within the broad psychiatric paradigm of the late nineteenth century (see Freud 1966), and contributed substantially to subsequent psychiatric conceptions of madness by way of his theoretical influences on early-twentieth-century psychiatrists such as Bleuler (Bleuler 1961).[17] The circle is closed, as it were, when the anthropological literature is used in turn by psychiatry in the theoretical 'transcultural' endeavour. Escaping the grip of psychiatric, or at least medical-scientistic, discourse when talking of madness today requires considerable effort.

A further complication in attempting to grasp Melanesian conceptions of madness is the lingering effect of the superficial facility of *lingua franca* terms adopted during the colonial period for convenience, and the assumptions held during the same period about Melanesian attitudes towards afflictions and physical disabilities. In Tokpisin, the main *lingua franca* of PNG, the term commonly used during the colonial period for craziness was *longlong*, and this became institutionalized in psychiatric and some anthropological literature as a kind of bridging term between the Western notion of 'insanity' or 'mental illness' and indigenous terms interpreted to mean the same thing. The country's most prominent psychiatrist, B. G. Burton-Bradley, wrote prolifically on the subject using *longlong* in this fashion (for example, 1973, 1975). But *longlong* does not correctly reflect traditional Kakoli conceptions of madness (nor, I suspect, those of many other Melanesian societies). Nor do more recent Tokpisin terms such as *waialus* ('loose wire') or *mentelkes* ('mental case') that borrow from Western, rather than Melanesian, idioms. Similarly inadequate, in its day, was the term from 'police Motu' (nowadays often called 'hiri Motu'), the *lingua franca* of the south PNG coast. Early colonial officers in the region adopted the term *kava kava*, which in 'pure' Motu (the language from which police Motu was derived) originally referred to wild or uncontrollable behaviour. Their usage of it (sometimes shortening it to '*kava*') paralleled that of the Tokpisin term *longlong* (see, for example, Monckton 1922: 153), as did Burton-Bradley's (see, for example, Burton-Bradley 1973a: 32–34[18]). The Motu term has more or less disappeared from popular usage in recent decades in favour of the Tokpisin term.

Derivative assumptions (ironically with folk-European precedents) about Melanesian attitudes toward disabilities include a belief that deaf or deaf-mute people are classified as mad. It is difficult to tell whether such assumptions have influenced anthropologists, for deaf-mute conditions are sometimes included ethnographically in discussion of madness. I cannot comment on the validity of, for example, Reay's (1965: 18–19) statement that among the Kuma (western highlands), the term *kumugl* used of deaf or deaf-mute people also meant 'mad' or 'crazy'. The same was certainly not true of the Kakoli. However, the conflation of 'deaf' and 'mad' is implied in a linguistic paper on Umbu Ungu (Blowers 1970). This is worthy of note because it indicates a confusion of the identification of *kekelepa* (social) behaviour itself with explanatory or causal statements about *kekelepa*. Blowers translates '/pimu.i/' – the interrogative use of the term *pimu* – as 'Are you crazy?' (1970: 4). The term *pimu* did not however mean 'crazy', but could be translated as 'blocked', as in cases where a passage or hole has been stopped up. It was used of people who were 'deaf-mute' or even both deaf-mute and blind, that is, people whose senses were inoperative to a degree that seriously impeded their awareness of what was happening around them and their ability to communicate. This was not in itself madness but it could of course be an explanation for strange or inappropriate behaviour, that is, *kekelepa* behaviour. The exasperated question 'Are you *pimu*?!' (for example put by parents to naughty or recalcitrant children) was rhetorical, offering a caustic explanation for *kekelepa* behaviour. The distinction between *kekelepa* as a way of behaving and the range of things which might be mooted as causes or explanations for it is important, and in the interests of clarification a comparative discussion of illness and *kekelepa* is required here.

Madness, Sickness, Causes and Cures

The Kakoli were amenable to a wide range of explanations for *kekelepa* behaviour and the rationality of these was related to the particular behaviour involved in any given episode. Alcohol was an accepted cause, as implied in my earlier note that drunken behaviour was glossed as *kekelepa* (although younger people frequently substituted the Tokpisin term for drunkenness, *spak*). Physiological explanations for *kekelepa* could be mooted, including physical injury, the ingestion of contaminated food or liquid, or even (as indicated in the case of Gawa in chapter 2) a severe headache. The moral lapses of a person could sometimes be seen as a cause of *kekelepa,* as in the case of a man whose mother posthumously punished him for his chronic neglect of her (particularly in her old age) by making him *kekelepa* for a period of several weeks (see Chapter 4). In another case, a widow who regularly declaimed nonsensically in public was said by some people to be suffering punishment from her late husband for ignoring his dying wish, marrying again quickly and into a different clan than his. Someone suggested alternatively that she had annoyed God by attending churches of too many different denominations.

The two latter explanations indicate that *kekelepa* could sometimes be explained by phenomena that show some kinship with Western ideas of 'possession'. Usefully, for comparative purposes, the influence of the dead is described in an ethnography of the Kumdi Engamoi (who display linguistic and other similarities to the Kakoli) which includes a brief passage on madness (Brandewie 1981: 177–78). Brandewie gives a short account of a young man who periodically ran 'amok', due to the spirit of his dead father entering his belly (ibid., 178). The account is given in a section on 'Shamanism' and types of divination, although no specialist divination was used (or appeared necessary) to explain the young man's affliction. In the context of my foregoing comments on interpretation and explanation, the terminology used in Brandewie's introduction to the episode is noteworthy:

> [T]here are people who periodically go *longlong* or temporarily out of their mind. This is called *kekelip ronum* as opposed to a permanent state of derangement called *kurkurpur ronum*. (ibid., 177)

We note (recalling my discussion in previous pages) Brandewie's use of the Tokpisin term *longlong,* and of 'out of their mind' and 'derangement'. Brandewie does not translate the terms *kekelip ronum* and *kurkurpur ronum* from Bo Ik (the language of the Kumdi Engamoi), but *kekelip* is clearly a dialectical relative of the Umbu Ungu term *kekelepa,* and *ronum* is from a root, shared by Umbu Ungu (*ro-*), from which some verbs of presence and transition are formed: *kekelip ronum* could be translated into English using a construction like 'become mad', or 'gone mad' or 'turned mad', and its equivalent in Umbu Ungu would be *kekelepa ronum.* The Bo Ik term which Brandewie renders as *kurkurpur* is possibly an equivalent of the Umbu Ungu *kurpule,* which referred to the theft of a person's *mini* by an invisible being such as a dead parent or a *kuru walu* (feral *kuru*), a malevolent entity that inhabited areas peripheral to human habitation.[19] The victim was left in a state either of behavioural vacuity or of frenzy (the term 'amok' would not be entirely inappropriate for the latter), depending on how his or her *mini* was treated by its captor. Finally, returning to my distinction between *kekelepa* itself and its causes, the description *kurkurpur ronum* (or *kurpule ronum* as Kakoli might say), refers to a cause (or possible cause) of *kekelepa,* not to *kekelepa* itself. I should add, *contra* Brandewie's statement about the permanency of *kurkurpur ronum,* that among the Kakoli, this was not necessarily so.

The Kakoli accepted that, on the one hand, a person could become permanently mad and, on the other hand, someone who had been mad for a significant length of time could suddenly become normal, without perceptible action or intervention by others. For example the manically declaiming woman mentioned above, who had been *kekelepa* when I began fieldwork, suddenly became normal – overnight – several months later. I was mystified, but the community seemed unsurprised and not at all intrigued by what seemed to me to be a most remarkable turn. When pressed, they surmised that maybe God or her husband had stopped punishing her, or she had made amends in some way, just as the *kekelepa* behaviour of the man who had neglected his mother came to an end after he killed a pig to make amends (for the dead as well as the living enjoy cooked pork). The formerly *kekelepa* woman herself expressed a lack of understanding ('*naa pilkiru*') of my questions when I broached the matter with her, though it was clear from her demeanour that she did not want to discuss it with me.

The varieties of explanations for *kekelepa* that I heard during fieldwork did not amount to a logics of causality, and it was partly for this reason that the local *alaye* (diviner/healer of illness) insisted to me that neither he nor anyone else could 'cure' *kekelepa* through intervention. The *alaye*, named Manenge, was a member of my host clan and was one of my key informants. Some elaboration of his role and activities in matters of 'illness' and 'curing' will help clarify Kakoli attitudes toward madness. Manenge had himself experienced an episode of *kekelepa* in his youth as a condition of acquiring his knowledge of healing, which he practiced through the application of prepared plant materials and secret incantations. During his *kekelepa* period, described for me by clanmates, he spent much of his time in the forest and roaming the upper slopes of Mt Giluwe (i.e., away from the usual areas inhabited or frequented by people), occasionally raiding other people's gardens for food. This 'theft' had been tolerated, as it was recognized that his madness was a contingency of his becoming an *alaye*. Informants described Manenge's speech as having changed during his *kekelepa* period, involving glossolalia ('he spoke many languages mixed up') and strange singing.[20] After a while he returned to normal, and with the guidance of his father (who was passing on his own skills patrilineally), cultivated the ability he now had to cure illness. Interestingly, the description of his overall transformation into an *alaye* adhered to the classic 'rites of passage' schema – separation, threshold, aggregation – proposed in 1909 by Van Gennep (1960), and elaborated particularly in terms of the threshold, or liminal, phase by Victor Turner (see, for example, 1969).

To be 'ill' was spoken of in Umbu Ungu as *kuru tokomo* ('*kuru* hits' or '*kuru* takes hold'). As I have already said, I cannot give a precise translation of *kuru*, a word used both of 'spirits' and of illness. My fieldworking predecessor Nancy Bowers regarded these two uses of *kuru* as homonymic,[21] but I was not convinced of this. Some years after my fieldwork I read the recently translated (1990) monograph by the German missionary Herman Strauss which was originally published in 1962, in which he addressed the notion of *kör* (the Melpa dialectical equivalent of *kuru*) at some length (Strauss 1990: 111–21). In this passage he described *kör* thus: 'For the Mbowamb … the collective term *kör* covers both the spirits of the dead and all the spirits which we would call spirits of growth or of nature, or else demons, according to their functions' (ibid., 111). Strauss's discussion gave strength to my disagreement with Bowers' understanding and indicated the complexity of the connotations of the term *kuru*. Whatever understanding I have of some Kakoli concepts to which I have already referred (including *numan* and *mini*) comes partly from extended discussions with *alaye* Manenge, who was locally recognized to have a particularly good knowledge of such matters. These discussions were hampered by my lack of fluency in Umbu Ungu, which Manenge tried to compensate for by frequent translation into Tokpisin (in which I was fluent from previous experience) and by enlisting assistant clansmen, of whom some younger ones spoke reasonable English. Manenge was very articulate and his exegesis was normally very confident (and, I must add, patient in the face of my own shortcomings). Interestingly, though, on the topic of *kuru,* he was hesitant, and commented that he did not think he could explain this to me well enough, and that there was no way of adequately translating *kuru* into Tokpisin. Other people I asked about *kuru* could only generalize about invisible presences or 'spirits'.

To be an *alaye* required not only a knowledge of types of illness and of the medicinal remedies (usually mixtures of a variety of plant materials) but also the acquisition of an enhanced communicative relationship with the natural world. This, in fact, was why Manenge had been temporarily *kekelepa* when he was young. His *kekelepa* period was part of a processual régime, gaining him extraordinary abilities to 'talk' with the Kaugel environment. To these abilities were added diagnostic and remedial skills normally learned patrilineally. Manenge explained this aspect of his practice after lengthy negotiations between us, for it involved 'secret' knowledge, which he would not normally divulge. The particular conversation in which he spoke of this knowledge was held in private: Only Manenge, his eldest son and myself were present. At the time Manenge was considering grooming the son to become his successor as the community *alaye,* and was assessing his maturity, demeanour and likelihood of commitment to the responsibility. During our conversation Manenge recited sections of two incantations to me as a demonstration of his persuasion of the plant materials to perform curative actions after they had been consumed by the sick person. His brief recitations were given on my undertaking never to repeat them, a promise which I have kept. I can, however, give a rough idea of the nature of the communication involved in such incantations.

The incantations were performed, for example, after the curative ingredients had been put into a bamboo container (Manenge would not use any other kind of container, including store-bought cups or mugs, for medicinal purposes) and before the container was given to the sick person. They were spoken *sotto voce,* and away from the hearing of the 'patient' and other people, and had a poetic and rhythmic form. Manenge addressed the ingredients both separately and collectively, and used allusions to naturally occurring phenomena and the local environment in his persuasions. For example an allusion to a local stream which had been made torrential by a storm and tore trees and boulders along its course from their positions could be used to indicate to the ingredients how they were to dislodge a foreign object clinging to the lining of a man's gut.

Complementing this vocal communication was the repertoire of incantations, including sung chants, employed by Manenge when collecting his plant materials, which involved a trip up the mountainside away from the inhabited area. In addition to the diagnosis of physical afflictions and the administration of cures he was often called on to divine the source of an illness suspected to be the result of human malevolence and perpetrated through sorcery or poisoning.[22] Illness was said also to result from unresolved or unrecognized social problems or tensions, and sufferers and their close kin would commonly try to puzzle out who it was that might be harbouring anger toward them, and *vice versa*.[23] It was recognized that not all ailments were caused by human agency, and simple remedies were applied to such things as minor headaches, toothache, coughs and colds without consultation. When Manenge was consulted by sufferers his methods of investigating the source of illness, selecting and preparing the appropriate medicinal plants, and applying the appropriate incantation were systematic, as was his divination in those cases where human malevolence was suspected of causing the illness. His methodical approach could be construed in terms of a rationality of corporeal illness and its treatment, amounting to a logics of causality, and therefore possible cure (and Manenge had a canny understanding of what he could and could not cure).

It was by virtue of this rationality that Manenge declared himself unable to 'cure' seriously *kekelepa* people. While explanations for *kekelepa* behaviour could be mooted, there was no tangible process in whose course Manenge could intervene or which he could manipulate therapeutically. Indeed, unaware of the way 'mental' and 'physical' illness are discursively related in Western medical science, he was at first unable to understand why I asked him, early in our acquaintanceship, whether he could 'cure' *kekelepa* people. Extended discussions with Manenge enabled me to discern the horizons of the traditional Kakoli conception of illness and disease in comparison to contemporary Western perceptions that include some forms of madness (defined as 'mental illness') in a medical-scientific paradigm of illness. Manenge, for his part, was bemused by my description of psychiatry and my explanation of its inclusion in medical science, but he came to understand my research into local madness better as a result of our discussions. Subsequently he was more forthcoming about his own healing practices, though guarded about precise details, as I have already indicated. He told me he had been suspicious of my motives in questioning him on those matters initially since I had explained that my research was focused on *kekelepa* people, which he had not seen as being relevant to his own occupation. When he discovered the reason for (in his view) my confusion of *kekelepa* with illness, he was keen to help me understand the difference between them.

During one conversation with Manenge, I queried his declaration that there was no 'cure' for *kekelepa*. I was referring at the time to florid and dramatic madness, of a degree that would in Western society invite psychiatric diagnosis of it as serious psychopathology, and I raised with him a local case that I was currently collecting information about. A man of my host clan had told me of how he had given a number of pigs, 800 Kina (a large sum in those times – a Kina is a PNG dollar), shells and food gifts to a variety of curers from distant places when his younger brother had become *kekelepa* several years previously. While none of the attempted cures had worked, an inference was that some Kakoli thought a cure was possible. Manenge, who knew about the case, commented that sending to distant places for cures demonstrated the man's concern and care for his brother, but that he had wasted his pigs and money: There simply was no 'cure'.

Manenge's point can be better understood with the aid of a short review of the case. The younger brother, whom I will call Ranje, had gone mad while working at a tea plantation near Mt Hagen. He had been brought back to the valley in a state of agitation and violence. During his more florid periods kinsmen tied him up and locked him inside a house. A total of eight pigs were said to have been sacrificed by his family in case his madness was due to the wrath of a recently deceased close relative. When this failed to bring about a change in his condition his elder brother took him to Mt Hagen hospital, but he was so uncooperative and hard to handle that nothing could be done with him. The elder brother then sought the services of curers from other areas, reducing himself, he told me, from relative resource wealth to poverty in the process. All these efforts failed to restore Ranje to normality. After a long period of time, however, the severe madness did wear off, but Ranje had been prone to occasional silliness ever since.

I spoke to Ranje himself on several occasions. His own versions of the onset of madness varied among themselves, apart from the fact that it had begun at the tea plantation, and the chronology of events was confused in his recollection. In our first conversation he said that he went crazy on a beer-drinking spree, but did not know what had made him crazy. He qualified this later by commenting that he had not thought he was crazy, but everyone else said he was; he had been tied up and transported back to the Kaugel Valley in a truck. In another conversation he said the madness had begun after he had been given an injection at a health centre. He had originally felt unwell, but after the injection he became violent and knocked someone down, after which he was tied up and sent home. Ranje told me he was alright now, but prone to slight craziness now and again. This could happen in a noisy crowd, or at a public ceremony, or perhaps at one of the *ambu kinan* – group courting sessions – periodically conducted for young people.[24] People would make fun of him on such occasions and he would feel ashamed, he said. He wondered whether he would go dramatically mad again at some future time, and this bothered him.

Two incidents involving Ranje came to my attention during fieldwork. One took place at an *ambu kinan* session one night (I was not in attendance). Two young men talking to me about the session the following day mentioned in passing that Ranje had been there, but somehow managed to tear a girl's clothing, so she tore his shirt in return and with the other girls threw him out of the house. Ranje, they added, often made a spectacle of himself at *ambu kinan* sessions. The other incident followed a visit to myself by Ranje to get cough medicine (I had a small first-aid and medical kit). When he left, a towel that had been lying on the grass outside a nearby house went missing. A few days later the owner of the towel saw Ranje wearing it as a loincloth and complained to the local *kaunsel* (Tokpisin: 'councillor' and settler of minor disputes – a survival from appointments of the late colonial period). The *kaunsel* suggested that Ranje should pay the woman K4 as well as returning the towel. The woman angrily announced that Ranje had better pay up smartly or she would take the matter to a higher authority. Several people commented to me afterwards that the woman was overreacting and being a little harsh on Ranje. After all, they said, his behaviour was often slightly *kekelepa:* Hitting him and reprimanding him would have been a more appropriate response, in their view. It was clear that Ranje had come to be regarded as more prone to social lapses than most people. Before his *kekelepa* period he had been responsible and smart, according to his peers.

I could achieve no more clarity about the circumstances of the onset of Ranje's madness. The degree to which his state had been exacerbated by the reaction of others was difficult to judge. Ranje himself told me once that their reactions had made him worse: 'People said, "this young man is mad", and the more they said I was mad, the madder I became'. The elder brother's efforts to help Ranje reveal, in their desperation and costliness, the truth of Manenge's point about *kekelepa:* In the Kaugel Valley there was no institutional 'cure' for it. Manenge's distinction between what could and could not be 'cured' might be usefully recast as a distinction between the kinds of afflictions which could be remedied by a direct and tangible engagement with their material cause (which is what his cures involved) and those kinds which could not be dealt with in such a way. Propitiating the dead, for example, by killing pigs in order to persuade them to end an affliction they had visited upon somebody, was not classified as 'curing'. *Kekelepa* behaviour was only sometimes a matter of affliction. It certainly was not affliction, for example, in the case of drunkenness or Manenge's own period of *kekelepa,* which was a necessary liminal episode in his acquisition of healing knowledge. So far as the influences of the dead were concerned, only an intimately related and recently dead entity such as a parent or spouse could have the kinds of effects on people that might result in seriously *kekelepa* behaviour (this will be discussed further in chapter 4). As I said earlier, Manenge was bemused by my description of psychiatry, and its discursive inclusion in the field of medical science. The reasons for his bemusement are understandable in view of the foregoing discussion and imply that the Kakoli would not interpret the intent of psychiatric services in the same way as laypeople in the West. I turn now to Kakoli attitudes toward psychiatry.

Psychiatry according to the Kakoli

I described in chapter 1 how psychiatry formally arrived in PNG in the late 1950s, after a period during which non-psychiatric doctors used portable electric shock machines and injected major tranquillizers to control Melanesian patients deemed to be insane. I noted that the treatment did not change significantly after the arrival in 1959 of B. G. Burton-Bradley, who worked energetically to extend psychiatry throughout the country and gradually established psychiatric wards in most hospitals. Psychiatric professionals regard their profession as the treatment of mental illness, and the accoutrements of psychiatric practice are their tools in this endeavour. The Kakoli, lacking a traditional subscription to a concept of mental illness, did not share this understanding of psychiatry. According to Kakoli sociality, seriously mad people were the responsibility of their kin. In fact during the colonial era socially unproblematic madpeople were concealed from European patrol officers who were empowered to remove them.[25] Crazy behaviour was tolerated, or even encouraged for amusement, but the tolerance diminished in the event of material or personal damage. If *kekelepa* individuals became too socially disruptive or their actions strained intra-group relations they were put under restraint by their close kin. During fieldwork I was told of past strategies such as securing disruptive madpeople in houses or tying them up under such circumstances. Apart from demands for compensation for damage caused, hostility toward madpeople abated when they ceased to be materially dangerous.

While binding and incarceration in a house might appear to be reasonable and straightforward short-term solutions to violent madness, they were often less than satisfactory for the community. Locking a madperson in a house invariably required people to act as 'guards'. I was told with retrospective humour by one man of how he had been obliged to stand guard for a considerable length of time when Ranje (see above) was locked in a house and raging incessantly. As the houses were constructed of bush material, there was no guarantee they could withstand a determined attempt to break out. Colonial intervention in this kind of containment in the highlands was probably initially unsystematic and reliant on the individual responses of *kiaps* assigned to rural areas. But the establishment of European medical facilities brought with it the use of sedative injections, exposing Melanesians to a simple and effective alternative to customary methods of controlling madness. The combination of authoritarian intervention by Europeans (employing police officers if necessary) and the use of medical officers in controlling behaviour which Europeans found undesirable provided indigenous communities with a new resource in dealing with disruptive individual behaviour. At the same time, expedient utilization of law officers and medical resources as a corporate control agent did not require any acceptance or understanding of medical science in general.

Papua New Guineans have never been driven to submit kinspeople to psychiatric facilities through any economic inability to support them, as has sometimes been the case in Western society. The colonial institutions simply represented alternative incarceration mechanisms to their own and were increasingly exploited to that end as post-war familiarity with European institutions grew. At the time of my fieldwork among the Kakoli, regulations had recently been introduced to ensure that a kinsperson was present during any patient's stay at the Goroka Hospital psychiatric ward. They failed to stop what the European staff called 'dumping' by the community. In the eyes of rural Papua New Guineans incarceration was still the major function of medical facilities in respect of madpeople, who were presented either directly or through the agency of the police, usually in the event of violence. Some examples of this will be given in a later chapter.

Insofar as the Kakoli understandings of madness discussed in this chapter are representative of a constellation of 'pre-psychiatric' worldviews in PNG, they should not be seen reductively as evidence of simple ignorance of the 'true' function of psychiatric services. Rather, the expedient attitude toward the constraint function of medical institutions was a contingency of the traditional pragmatic management of madpeople in such communities. The development of psychiatric facilities at regional hospitals in the late 1960s served to reinforce the indigenous view: In the highlands the 'open' wards at Mendi, Mt Hagen and Kundiawa (Simbu Province) were generally unable to contain the violently manic individuals taken there by the police or the community. Invariably the patients were sedated and shifted to Goroka Hospital, where at the time of my fieldwork the psychiatric ward still consisted of a cage-like holding cell and a corridor of single rooms with steel doors.

In the late 1970s Goroka Hospital staff attempted a follow-up survey of ex-psychiatric patients, after which the author of the report warned against accepting the reliability of solicited community opinion on the efficacy of the treatment, commenting that the data might reflect '[a]cquiescent-set responses by villagers to the people conducting the interviews (hospital staff members) whom the village people would not like to alienate (or would like to give a favourable impression) so that the *longlong* person, should he/she become disruptive in future, could again be removed to that available and convenient resource known as hospital' (Robin 1979: 178).

Lacking historical conditioning into the acceptance of medical scientistic paradigms, Melanesians have tended to judge the medical scientific treatment of madness purely on its ability to return individuals to social normality. This attitude contrasts with that of laypeople in Western society, whose acceptance of psychiatry as institutive of social control has been achieved ideologically via their capitulation to authoritative discourse about mental illness, through which any failure of psychiatry to effect cures is rationalized. With the advent of medication enabling psychiatrically diagnosed people in the West to survive extramurally with the aid of 'halfway houses' and social workers, blame is shifted further from the hospital: if extramural 'patients' occasionally become disruptive or violent, the fault is seen to lie in the patients' mishandling of medication or the community's inability to become more involved in mental health care.

In comparison, the introduction of post-discharge medication in rural PNG during the 1970s served to demonstrate the futility of psychiatry as a therapeutic force. The patient usually stopped taking the medication within days, and the subsequent discovery of still-full containers of pills by medics or follow-up workers months or even years later is medical lore and has been documented (especially for the Simbu and Eastern Highlands Provinces) by Robin (1979: 123–29). The phenomenon tended to be approached by professionals as a problem of making sure people understood that they were to take the medication regularly and of finding ways to help them remember to do so. Superficially this was a reasonable interpretation, and efforts to ensure regular pill-taking were hampered by the indigenous lack of the minor systematic rituals into which Western pill-takers articulate their self-medication (for example, bathroom and bedtime rituals).[26] But the medical ignorance implied by such an interpretation was not the simple cause of laxity in pill-taking.

When the patient left the psychiatric ward the violent symptoms had gone, removing the major reason for the psychiatric encounter. The administration of post-release medication relied on a particular ideology of madness: the assumption that the patient had a mental illness which required chemical intervention to keep it under control in the future. This was not coherent to peoples such as the Kakoli. If the recipient did not subscribe to the medical-scientistic paradigm, the medication was conceptually reduced to something judged on its brute effects. The drug given to patients on release in PNG was (and still is) almost exclusively chlorpromazine, which is a major tranquillizer. It generates a 'woolly-headed' feeling in those who take it, and can produce a variety of side effects, including weight gain, menstrual and lactation problems, Parkinsonian reactions and sexual dysfunction. In the West these are often countered by a secondary medication.

The Kakoli were not inclined to continue taking medication when debilitating effects were experienced, and if the socially disruptive behaviour had disappeared there was no community pressure on the individual to continue with the dose. A converse example of the latter is provided by the case of Gawa, the severe headache sufferer who was discussed in chapter 2. Gawa was released from Mendi hospital on a régime of chlorpromazine after being diagnosed as schizophrenic. Part of his subsequent discomfort, which he discussed with me, was caused by community interpretations of the effect of the drug on his demeanour. His medicated behaviour was in itself regarded as *kekelepa*. Judged purely on its ability to return individuals to social normality, then, psychiatric medication could achieve little from the point of view of a society like the Kakoli, where 'welfare' support services did not exist and people were dependent on their normal faculties for their subsistence and social survival.

Local interpretations of psychiatry contributed to misunderstanding between psychiatric staff and the kinsfolk who delivered a *kekelepa* person into their care.

As we saw in chapter 2, for example, once the severely disruptive or violent aspects of the person's behaviour were controlled by medication, the kinsfolk were likely to want the patient discharged, regardless of any other 'delusional' or verbal aspects which the psychiatrist might regard as evidence of a need for ongoing treatment. Conversely, though, people who had been found chronically unmanageable by the community might be left indefinitely at a psychiatric facility if it were thought there would be no discernible role for them in communal life in the future. European psychiatric staff with whom I spoke in the course of my research referred to this as 'dumping', a word whose inference I considered inappropriate to the sentiments of the community. I saw no evidence, during fieldwork, of callous attitudes by people toward chronically mad kin, and there was in fact a reluctance to relinquish kin entirely to psychiatric or other incarcerative institutions.

The reductive Kakoli understanding of psychiatry – that it is a social control resource – is consonant with the absence of psychiatrism, and its necessary correlate psychologism, in their own view of madness. The exegesis in this chapter has rigorously avoided any psychologization of madness in the Kakoli lifeworld, and has explicitly denied the validity of applying a psychiatric paradigm. I have shown that the rationality that the Kakoli applied to considerations of physical illness, amounting to a logics of causality, did not extend to considerations of madness. This is because madness was not classified as an illness, and certainly not as a 'mental' illness, and was therefore not drawn into a category amounting to an indigenous equivalent of medical science.

To summarize some main points of this chapter: Psychiatric understandings of madness involve a perception that it is intrinsically 'mental' – that it is in the mind-in-the-brain – even when 'psychosomatic' processes are acknowledged (since psychosomatic considerations presupposes the presence of the psychological aspect). The projection of psychiatric definitions into the Kakoli lifeworld necessarily applies a psychological individualism to which the Kakoli did not subscribe. The mind (in-the-brain) was not a psychological locus of individualism among the Kakoli, who subscribed instead to a contingent relationship between thinking (*konopu*) and the socially developed, and socially potent, *numan*. The *numan*, as invisible as the Western 'mind', was located in the chest rather than in the brain, and was actively an intersubjective force, rather than a 'thinking' entity. Further, the Kakoli definition of madness as *kekelepa*, behaviour that is *socially* out of place, admits of a variety of explanations and 'causes', with no special emphasis on the psychology or 'mind' of the *kekelepa* individual. Any sincere attempt to grasp the phenomenology informing Kakoli ideas about madness must therefore forgo psychologism. This thematic imperative will guide the ethnographic material in the following chapters.

Notes

1. This is also subject to various spellings in academic and non-academic literature (for example, Umbugu, Umbungu, Umbu Ugu). The word *Umbu* is from a root denoting autochthony, and could be translated by 'local', 'natal', 'seedling' or other glosses expressing organic genesis: *Ungu* can be translated as (the noun) 'talk'.

2. The word *mbowamb* draws together some linguistically similar terms in the area. Strauss (speaking ethnologically of the area as a whole) translates *mbo* as 'seedling, cutting' and *wamb* as a combination of *wö* (man) and *amb* (woman), translating the compound as 'seedling people, cutting people' (1990: 2). Brandewie (speaking ethnographically for the Kumdi Engamoi) similarly translates *mbo* as meaning 'seedling' or 'runner' (as of sweet potatoes or trees), and *wamb*, as a combination of *wua* ('man') and *amp* ('woman') (1981: 32).

3. As I have implied here, Bowers's thesis was about much more than sweet potato, and her considerations of historical demography took into account warfare, climatic conditions and changes, oral traditions, and so on. She published little of the substance of her thesis, but a summary of sorts is provided in a 1971 paper (Bowers 1971: 11–31).

4. Two kinds of 'traditional' story were distinguished by the Kakoli. These are *temane* and *kange*. I do not agree with Bowers's representation of the difference between them, that a *temane* is thought to be 'true' whereas a *kange* is 'just a story' (Lepi and Bowers 1983: 4). I am inclined to see *temane* as stories intended to explain directly or indirectly some aspect of the present by narratives of past events, and *kange* as stories whose narrative content does not particularly have that intention, and which are told simply for entertainment.

5. A little Kaugel Valley oral history has been published (Head 1974; Karoma 1978; Didi 1979; Lepi and Bowers 1983). The best collection is *Kaugel Stories: Temane and Kange,* edited by Lepi and Bowers (1983). In relation to sweet potato, this collection contains one nicely exemplary story specifically concerned with the arrival of three clans in the valley (1983: 140–45). There is still debate about the antiquity of pigs in PNG (see, for example, Hide 2003: 12–13), but they definitely arrived before the sweet potato.

6. These official census-based figures are probably exaggerated, as a great deal of replication has become common in post-colonial census data. Also censuses were by now including part of the upper Nebilyer population in the Tambul figures. I suspect the population of the Kaugel was less than 28,000, and more in the region of 20,000, but I did not have the resources to test the official statistics. Census data were more reliable in Bowers's period (see Bowers 1971: 3–4).

7. N. Bowers, pers. comm., 1986. 'First contact' stories offered by Papua New Guineans nowadays frequently contain a statement to the effect that Europeans were thought to be ghosts of the local group's ancestors. This has become a popular refrain and is not to be trusted.

8. District Commissioner R. Y. Skinner, correspondence to Western Highlands Assistant District Officer, Mt Hagen, 10 June 1959, Subject: 'Patrol report, Tambul No. 1 – 1958/59'. PNG National Archive, Port Moresby.

9. 'Spirit' is the most common translation used by English-speaking Western Highlanders and anthropologists who have worked in the region. See, for example, Bowers (1968); Brandewie (1981: 182–83); Didi (1979).

10. In Umbu Ungu, when followed by a front vowel (/e/, /i/), as in *kekelepa,* the /l/ is pronounced as a laterally released dental plosive: to a native English speaker this would sound

like /dl/. When followed by a back vowel (/a/, /u/), it is pronounced as a laterally released velar plosive (see appendix A).

11. As with a number of other terms, the 'meaning' of *kelepa* was changing during my field-work. Linguistic changes in the Kaugel Valley included the increasing adoption of Tok-pisin words and the extension or alteration of the usage of Umbu Ungu to accommodate imported material goods and ideas. An important, but variously effective, influence on altered meanings (or at least changed usage) was Christianity, using Umbu Ungu verbally in church services and in written form in Christian literature produced for increasingly literate young people (see my concluding chapter for further discussion). In this kind of literature, *kelepa* was used to mean simply 'leave' or 'depart from'. Finally, *kelepa* does not mean to 'go', or to 'go away', for which the root is /pu/.

12. From the root *ke-* comes, for example, the word *kel* meaning 'small'. However, there is a difference between the sense of 'small' expressed by *kel* and the sense of 'small' expressed by another word, *walo*. The latter term expresses a static sense of size, as in *ye walo*, 'small man'. *Kel*, on the other hand, has a comparative sense of 'diminished', or 'less'. A small boy might be referred to as *walo* (*kango walo* = small boy). Alternatively, though, he could be referred to as *kel* (*kango kel* = small boy) whereby the use of *kel* contains an implication that he has the potential to be bigger (that he is a 'smaller-than-he-will-be' boy, so to speak). Also, if someone is attempting something with too much effort, or force, an observer might exhort '*Kel! kel!*', meaning 'Take it easy!', or more literally 'Do small!'. Mothers murmured '*kel, kel*' (equivalent to the English 'there, there', or 'shhhh') as they comforted crying babies.

13. I will return to the Christian use of '*mini*' for 'soul' in the concluding chapter.

14. Kakoli concepts of *mini* and *numan* were not dissimilar (given some dialectical variations) to those of nearby peoples, such as the Melpa (see, for example, Stewart and Strathern 2001: 113–36; Strathern 1981; Strauss 1990: 100–03), the Ku Waru people of the upper Nebilyer Valley (see Merlan and Rumsey 1991: 226–28) and the Kumdi Engamoi (see Brandewie 1981: 181). Ethnographic variations in the explanations of these concepts may reflect difficulties of *verstehen*, and varying effects of Christianity among them (see, for example, Stewart and Strathern 2001: 115–25; and conclusion, this volume).

15. There is extensive literature covering these issues both separately and collectively: See, for example, Bromberg (1975); Bynum (1981); Doerner (1981); Foucault (1982, 2006); Ginzburg (1983); Levack (1987); MacDonald (1979); Macfarlane (1970); Mandrou (1979); Neugebauer (1981); Rosen (1968).

16. This development proceeded at different rates in England, France and Germany from the eighteenth century onward (Doerner 1981) but the English example, which Scull (1979) explores, is representative of the basic shared elements.

17. What is nowadays called 'schizophrenia' was, at the beginning of the twentieth century, mostly called 'dementia praecox' in Europe (see Kraepelin 1919). Bleuler coined the term 'schizophrenia' as an alternative in 1911 'because … the "splitting" of the different psychic functions is one of its most important characteristics' (Bleuler 1961: 8). The focus on the 'psychic functions' was at the time a new turn, influenced by Freud's emphasis on the relationship between conscious and unconscious processes and the analytic accessibility of the latter. For Bleuler this held the possibility of therapy (1961: 476–77) where previously containment and control of symptoms had been the dominant psychiatric response.

18. Burton-Bradley uses the short form, *kava*, claiming it is more correct. The root term *kava* refers to emptiness, and its contextual application can therefore refer to things empty

of material content and metaphorically to behaviour empty of 'purpose' or 'meaning'. Burton-Bradley's preference is understandable, then, but slightly at odds with what I have found in my own experience of the Motu language (both in archival material and recent fieldwork in a Motu village), where *kava kava* is the common usage. There are a number of errors in Burton-Bradley's representations of Motu and other language terms in his *oeuvre*.

19. Strathern describes something which seems very similar among the Melpa, referring to it as *kupörl* (Strathern 1981: 302; cf. 1975a : 349–50) .

20. Strauss describes a comparable, but not entirely similar, Mbowamb process in which a young man is 'taken and transported into ecstasy ... becomes frenzied, goes out of his mind' (1990: 123).

21. Pers. comm.

22. People were careful to know where food had come from, and men especially were fearful of poisoning via food. There were said to be no sorcerers in the valley, but they could be hired from elsewhere, such as Ialibu to the southwest.

23. This returns us inferentially to the nature of the socially attuned *numan*, through which internal frustration and anger (*popolu*) can be transmitted.

24. Young unmarried men and women paired off and sat side by side and cheek to cheek, singing. These sessions were common in highland societies, and came to be generalized by the Tokpisin term *tanim het* ('turning head'). A variant, in which the pair's legs were intertwined, was known as *karim lek* ('carrying leg').

25. Pers. comm., Nancy Bowers, 1987.

26. Mealtimes, or leaving for gardening, etc., were common mnemonic devices used by health workers. The problem was discussed with me by Brother Andrew, pers. comm.

4

AFFLICTION AND MADNESS

In this and the following chapter I describe some episodes of *kekelepa* among the Kakoli that were more than momentary and minor breaches of Kakoli sociality. I have already made a distinction between *kekelepa* itself, which for the Kakoli was a behavioural estrangement, and the discourse of its causes and explanations (and here I appear to differ from Brandewie [1981: 77–78] and perhaps some other anthropologists of highlands PNG). The discourse did not amount to a logics of causality, but was categorically wide-ranging and open to speculation about physical and non-physical, and ordinary and extra-ordinary, factors. It is noticeable, though, that among the explanations were some which attributed the madness to an affliction by the dead or by invisible entities such as *kuruwalu*, and in this chapter I will elaborate two examples of this type of affliction. They would undoubtedly have been diagnosed as forms of mental illness had they been brought to the attention of a psychiatrist, and psychiatric diagnosis would probably have acknowledged significant 'cultural' factors. However, rather than conceiving individual madness as primarily a psychological aberration which might or might not be mitigated or explained by its cultural context, I take account of the historical social context according to which the Kakoli themselves identified people as *kekelepa*, for their discourses of *kekelepa* behaviour which are at first sight stories of madpeople are also stories of themselves, their experiences and their preoccupations. This will become clear in the two descriptions given here.

I have already described the Kaugel Valley as an environment that does not appear inviting of human habitation. Indeed, this was a theme that drove Bowers's original study of Kakoli adaptation to the biosphere and her detailed description of their agriculture (1968). Enormous effort was invested in subsistence activities and remarkable transformations had occurred in the landscape by the mid twentieth century as a result. Conventional classification of the Kakoli as principally subsistence agriculturalists would be perfectly justified. They could not, of course, be imagined to be unaffected by the capitalist economy, even at the time of Bowers' study. Like many other Papua New Guineans, by the 1960s they had acquired a degree of capitalist ideation through experience of wage labour and profit enterprise. However, relations of production and exchange were far from being individuated to the point where group identity became undermined. Extended kin groups remained the major units of generalized production, consumption and exchange, and engendered an ideation of group responsibility for the action of individual members and group entitlement to compensation for individual misfortune at the hands of others.

The harsh environment of the upper Kaugel Valley required particular gardening skills and industriousness and a strong attendant sense of communalism, the latter in tension with both individual and clan jealousies and covetousness of good gardening land (from which many intra-valley wars developed before colonial pacification). While the socio-economic activities and attitudes of the community had changed under the impact of colonial administration, missionization and the cash economy, communalism remained an overwhelming ethos, manifest in fierce descent group loyalty and a rigorous adherence to obligations to share resources and reciprocate gifts and assistance. Conflict arising from jealousy over land, for instance, was restrained most of the time under this ethos. Community preoccupation with themes of social obligation and sharing is exemplified in the Kaugel Valley stories collected by Lepi and Bowers (1983), and was emphasized to me anecdotally during fieldwork. Those who defaulted on such obligations suffered at least a withdrawal of co-operation from their fellows, and more likely censure and punishment, sometimes direct and sometimes subtle. The pervasive complex of spoken and unspoken obligations impressed itself immediately on me during fieldwork, but I was also given a dramatic anecdotal example of the consequences of neglecting one's responsibilities shortly after my arrival in the valley. I was told by a number of people of an episode of *kekelepa* involving a man whom I will call 'Kapiye' which, they said, I would have been able to witness for myself had I arrived in the valley a few weeks sooner.

Kapiye: The Birdman of Kiripia

The story of Kapiye's madness was told to me not only as a response to my enquiries about *kekelepa* people, but also to emphasize the importance for the Kakoli of looking after one's immediate kin, and especially the duties of males toward ageing parents. Kapiye was a young married man of the Sipaka clan, who were the immediate down-valley neighbours and traditional war allies of my hosts the Kepaka clan. It was unanimously agreed that his recently dead mother had posthumously afflicted him in response to his neglect of her over a long period of time. Kakoli ancestors and the dead could afflict the living in a number of ways, causing personal illness, ruining crops, bringing disease to pigs, causing physical harm (of the kind Westerners would regard as an accident). The *mini* of a just-deceased person was extremely powerful and remained so for several months, and the living were cautious not to offend him or her for fear of personal attack. Those people most intimately connected with the deceased were the most vulnerable. As the months passed subsequent rituals ensured that the *mini* joined the phalanx of ancestors (*tarama*), at which stage it ceased to be regarded as a *mini* and became less individually feared. The period of time between death and the *mini*'s incorporation into the *tarama* was equivalent to that required (in former times, before Christian burial practices were adopted) for the corpse's soft tissue (*kangi*) to rot away completely, leaving only bones (*ombel*).

Kakoli ancestors and the dead could cause afflictions of many kinds, but they were not commonly able to cause individual madness directly. There were, however, exceptions. People could be afflicted with *kekelepa* behaviour by the *mini* of recently dead people with whom they had had particularly intimate relationships, such as parents or spouses. These were the relationships which traditionally involved the most stringent and transparent reciprocal responsibility and obligation. Children were said to be formed from the *meme* (blood) of their mother and the *kopong* (semen) of their father and to inherit their *mini* from ancestors through their parents; thus there was a powerful link between the *mini* of a parent and that of a child. In the case of spouses, the two *mini* of the pair were said to have become intimate to the degree where they could remain in contact after death. Lapses in the obligations between these intimately linked people left the living extremely vulnerable to the *popolu* ('frustration anger' or 'resentment', see below) carried by the dead person's *mini*. In addition to other afflictive action, the *mini* of the dead person could take control of the *mini* of the living person. This principle was implicit in an episode mentioned in chapter 3, in which a widow had gone mad after remarrying against her dead husband's wish. She should not have remarried until the cycle of post-mortem rituals was finished, by which time her husband's *mini* would have completed the passage to membership of the ancestral collective, or *tarama*. His displeasure might thereafter have manifested itself in some way, but not in the same direct and dramatic manner. Her hasty marriage, while his *mini* was proximate, resulted in his directly punishment of her by manipulating her *mini* and making her behave crazily.

The control of the *mini* of the living by the *mini* of a recently dead intimate relative was very evident in the example of Kapiye's madness. There is little 'domestic' privacy among the Kakoli, and Kapiye's failure to look after his mother was well known. He had gone away to work at a plantation near Mt Hagen and had not paid visits to the valley to see his ageing parents; and he had contributed nothing to their welfare. One elderly man told me that he had briefly been in the bare room that served as a ward at the health centre at Tambul at the same time as Kapiye's mother during an illness shortly before her death. She had no blankets, he recalled, and he had asked her why her son (who was now back from the plantation) did not visit her and bring her a blanket. She had complained that Kapiye always neglected her and said, according to the elderly man, that she would die soon and afterward Kapiye would suffer. Other people told me that Kapiye was known to hide food when his parents came to his house so he would not be obliged to share it with them.

There were two versions of an event that took place just before Kapiye's mother died, when an example of Kapiye's meanness was exposed to his mother by his own small daughter. According to one version, Kapiye had cooked the meat of a bird and shared this delicacy with his wife and child but hid it from his mother. The little girl did not eat her portion entirely, however, and took some to her grandmother (i.e. Kapiye's mother). The grandmother, on realizing Kapiye's deception, was heard to say that Kapiye would be reminded of the bird incident after her death, according to this version. The other version alleged that Kapiye was cooking some bird's eggs (also a delicacy) when his mother came into his house. The ailing woman, whose sight was poor, asked what he was cooking and he replied that it was just some small potatoes.[1] Later, however, Kapiye's daughter exposed the deception by taking an egg to her grandmother. In this version, like the other, the dying woman made an ominous remark about Kapiye and birds. The bird/egg incident was said to be the last act of meanness before Kapiye's mother died, which was regarded as significant in regard to what followed.

Shortly after his mother's death, Kapiye was to be seen at dusk sitting in the tops of tall trees, on branches that ordinarily would not have been strong enough to support his weight. There was no doubt among witnesses that his mother was responsible. Kapiye would fly from treetop to treetop. In the gathering darkness it became difficult for observers on the ground to track his movements: He would appear in the top of one tree, then be momentarily lost to sight before appearing in the top of another. He would spend the whole night thus, being finally returned to the ground at dawn. Sometimes his mother would take him, via the treetops, high up Mt Giluwe overnight, returning him in the mornings. The journey up Mt Giluwe could be verified by the plant scraps on Kapiye's person when he arrived back – the plants were only found high on the mountain and did not grow in the inhabited area.[2] Attempts to communicate with him while he was in the trees were met with bird-like whistling instead of words (witnesses demonstrated this for me by whistling repeated notes of equal length at a constant pitch).

In a short while (days, it seems) Kapiye was spending time in the treetops during the day as well as at night. In addition he became violent, chasing people and rampaging in gardens, causing a great deal of damage. He also decorated himself in crazy ways and talked gibberish. The tree-sitting and mania continued for several weeks, until Kapiye (in a period of lucidity) slaughtered and cooked a pig (i.e. sacrificed a pig to his mother). At the time of the killing he dressed in ceremonial attire and decorated himself carefully; everyone who had seen him agreed that he looked very fine indeed. After killing the pig he was no longer taken up into the trees and his self-decoration, rampaging and odd speech wore off. By the time I arrived in the valley people considered Kapiye to be almost completely normal again.

The above account represents what could be called the 'consensus version' of events. That is to say, the elements of it were common to all accounts of what happened, apart from the two variant descriptions of the final act of unkindness to Kapiye's mother. Individuals, however, added various elaborations, some of them extremely dramatic. I was told, for instance, that a group of young men locked themselves in a house with Kapiye one evening in an attempt to stop him from being taken into the trees. When darkness fell a great force pulled Kapiye toward the tiny holes in the walls (big enough only to admit the inevitable family of rats into the home). It took the combined strength of eight young men to restrain Kapiye. Furthermore, when they held him down he gave off a smell like a rotting corpse. The smell transferred to the skin of his captors. In the end, exhausted and almost overcome by the stench, they opened the door and let Kapiye go. The corpse smell clung to them until they washed it off in the river.

Another elaboration (offered by several people) was that during his madness Kapiye had the power to enter locked houses at night (i.e. through rat holes, etc.). One woman said Kapiye would leave things in the houses he entered in this way and come and ask on the following day for whatever he had left (her example was his machete). Sure enough, the house-dweller would find the object inside, with no explanation for how it had got there other than through a nocturnal visit by Kapiye. Some people claimed to have seen Kapiye hanging upside down from high tree branches and maintained that he descended from trees head first, like a lizard, and according to one person Kapiye once escaped mysteriously from a locked house, leaving no trace of his exit.

Kapiye was pointed out to me at Kiripia one day shortly after my arrival in the valley. He was slightly decorated with fern fronds in his hair and a little paint on his face, in the manner of young men trying to attract women admirers. My companions of the moment commented that it was hard to tell whether he was completely recovered from his madness or not. Word spread quickly that I was interested in talking to Kapiye and hearing his recollections of his experiences, and a couple of weeks after I had seen him at Kiripia he was brought to my house by a group of young men with the news that he was willing to talk about his *kekelepa* experiences. Kapiye was now undecorated, and seemed a fairly dour individual. The whole group crowded into my house to hear the story, and with Kapiye's agreement I set up my tape recorder. He was fluent in Tokpisin, but said he would prefer to talk in Umbu Ungu about such a profound experience. Acknowledging my inadequacy in that language, one of the group appointed himself translator into Tokpisin, and Kapiye's story subsequently went on tape in both languages. Comparison of the Umbu Ungu and Tokpisin version of events was to prove interesting later.

Kapiye's starting point was after his mother's death, thereby avoiding all mention of his alleged mistreatment of her. One evening, he said, he had been cooking some rice and had briefly reflected on the fact that there was more food for his household to share now that his mother was dead. This momentary discourtesy to his dead mother was responsible for all that followed, he thought. After the meal he had gone outside to the toilet, and a sudden wind-like force lifted him into the trees. He called out, but he was being moved from tree to tree and when people came out of their houses in response to his cries they were unable to retrieve him. He was moved from tree to tree all night and was returned to the ground in the morning. This happened each night subsequently. Kin tried to restrain him in the evenings, and at such times he heard noises in his head, such as whistling and a sound like the roar of a car engine. Some nights he was taken up Mt Giluwe, and on one such occasion he was thrown down from the top of a very tall tree, but caught just before he hit the ground.

In the daytime, he said, he would roam around the gardens, he would fight people, and he would sometimes sleep in the graveyard.[3] At one point he started making preparations to cook his wife and child, and his wife ran away. Sometimes he could see the forces carrying him in the trees, a collection of dismembered people – single arms, or legs, or heads – all working together. Sometimes they tormented him without taking him up into the trees. He would try to hide from them in long grass, unsuccessfully. They would offer him food, but he refused to eat. Eventually he made a pig sacrifice and then his mother released him.

Kapiye's story is given in its briefest form here. It was delivered a statement or two at a time, translated as he went by the self-appointed young man. At the outset I tried to coach my translator to mimic the first-person delivery of Kapiye to ensure the best representation in Tokpisin: however, he slipped chronically into a third-person rendition that I eventually deferred to in the interests of discursive continuity. There were many embellishments to the account given above. One was that Kapiye would sometimes be 'invisibly' present in people's houses (implying that people could sense the proximity of his *mini*). Another was that he slashed down sugar cane for no purpose. He was said to have killed a pig and eaten its flesh uncooked. This was a behaviour commonly attributed anecdotally to seriously *kekelepa* people, and it was one of the examples given to me when I first asked Kakoli what kind of behaviour they would consider to be seriously mad. It was a powerful trope of social estrangement, as pigs were customarily killed to feed others, not oneself, and they were always cooked in an earth oven, which ensured that ancestors were being fed as well. To kill a pig and eat its raw flesh would be to traduce fundamental social principles. Kapiye was also said to have worn red cordyline leaves. Green cordyline was commonly worn by men to cover their buttocks. Red leaves (connoting blood, danger, death) were worn for certain ancestor-directed ceremonies. If they were worn on other occasions they demonstrated an intent to kill someone. In Kapiye's case the wearing of *kaila* (red cordyline) illustrated his craziness and perhaps the potency of his dead mother.

Several months later I listened to the tape of Kapiye's testimony again, by which time my understanding of Umbu Ungu had improved. I noticed that the translator had added a number of points to Kapiye's story, including all the embellishments mentioned in the previous paragraph. Kapiye, being fluent in Tokpisin, must have heard these additions at the time, but made no effort to stop or correct them: nor had the other people present. Either Kapiye did not trust his own memory of events enough to question the translation, or he concurred with the translator's addition of points that he himself had overlooked, or he was happy to allow imaginative additions in the interests of an exciting narrative. The last possibility is not at all unlikely – the Kakoli placed value on good story-telling.

There was a marked contrast, however, between this *laissez faire* and Kapiye's refusal to be drawn on events that had occurred before his mother's death. Diplomacy prevented me from actually confronting Kapiye with the allegations I had heard, but I tried some indirect questions. For example I asked whether he had worked on a plantation near Mt Hagen. This, like other questions, was fobbed off quickly on the ground that it was irrelevant to the story Kapiye had come to tell, since it was before his madness and before the death of his mother. In due course Kapiye left, with those members of the group who were his clanmates. The Kepaka men stayed to talk, and I raised the matter of Kapiye's contention that his momentary reflections on the effect his mother's death had on food supplies had been responsible for all that had followed. This, they said, was untrue: Kapiye was too ashamed to tell me how he had mistreated his mother, but his conduct had been widely observed by the community. Everybody to whom I later related Kapiye's explanation rejected it as an evasion of the well-known truth. Some people even went so far as to assert that Kapiye's mother would have told him directly that she would punish him after her death. Important connections between emotions and sociality are implicit here, and some elaboration of Kakoli concepts of emotions is necessary.

A commonly used Tokpisin term used in describing emotional responses among Melanesians is *sem*, transliterated from the English word 'shame'. In PNG 'shame' is popularly spoken of as a powerful emotion, felt more strongly and caused in more complex ways than it is among Europeans. Burton-Bradley, for example, incorporated the concept into a discussion of causes of suicide (1973a: 92–94). The implication, though, that Melanesian 'shame' is an exaggerated version of European or Western 'shame' is misleading, and partly a product of the inadequacy of the English term and its Tokpisin equivalent. A range of emotions is collapsed together by the convenient terms *sem* or 'shame', including mild embarrassment, remorse, inferiority, awe, angst and also the extreme sense of humiliation and despair which can result in suicide.

A general Umbu Ungu term that lends itself to translation as 'shame' in this broad sense is *pipili*. But some important qualifications need to be applied. As I indicated in chapter 3, Kakoli emotions were situated and experienced bodily. They could sometimes, but not always, be visible to others through external bodily conditions (trembling, sweating, etc.), but importantly they were experienced within a person's torso. *Pipili* was located in the belly (*olo*), the same place as *popolu*, which was 'frustration anger', or 'resentment', of a kind usually concealed from public view. Also in the belly was *mumindili*, a 'rising anger' that emerged publicly in the form of *iri*. The term *iri* was used of rebukement, beratement, and furious tirades, which in contrast to *popolu* were expressions of unconcealed anger. In its more extreme form *iri* generated from *mumindili* was also the expressed fury of warriors. *Pipili, popolu* and *mumindili* (and its expression, *iri*) were related, as their shared corporeal position implies, and one could turn into another. Sometimes it was difficult to distinguish between them, as in some cases of suicide by someone experiencing great *pipili*. This variety of suicide was sometimes said to be vengeful (i.e. from *popolu*) in that it was intended to cause *pipili* in the person who had made the suicidal person *pipili* in the first place. A different and complex *pipili/popolu* episode occurred during my fieldwork when an ageing man responded to the abandonment of himself and other kin by his eldest son, who had moved to a coastal town and was leading a profligate life. In an act of self-mutilation that was an associative rejection of his *kopong* connection to his son, the old man cut off his own penis. A less dramatic and more common example of the connection between these belly-situated emotions was when *mumindili* – whose expression as furious *iri* was often accompanied by physical violence – became *pipili* after the violence was done (particularly if the violence was toward relatives, friends or allies). Two other belly-related terms also used to refer to *pipili* were *olo tekemo* ('belly-doing') or the short-form *olie*.

The concept *pipili*, then, should be interpreted as referring to various sensations experienced in the belly, which enables us to better understand why its emotional range was considerable and would be inadequately glossed by the English term 'shame'. Inasmuch as *pipili* was an activity in the belly it could, for instance, have an aspect experienced as a significant heaviness, or pressure, which was a kind of foreboding or angst. This should not be confused with simple fear, for which the Umbu Ungu term is *mundemong* ('heart-trouble'),[4] an emotion located higher in the torso and experienced as tension within the ribcage. *Mundemong* was the immediate fear of tangible things, such as a falling rock or tree branch, or a truck skidding on a dirt road. The combination of feelings of inadequacy, inferiority and angst, a major constellation of *pipili*, was of a different order to *mundemong*, and related to a significantly disturbed *numan*. As I said in chapter 3, the *numan* was a locus of thought, intention and will, able to affect and be affected by others and by relations with others, and contingent to the *mini*.

Returning to the case of Kapiye, his failure to look after his ageing mother was a very unusual situation, for most people would have been overcome by *pipili* long before the neglected parent died. His mother harboured frustration-anger, *popolu* (not potentially explosive *mumindili*), as was suggested by the old man's story of the encounter in the medical ward. But enfeebled as she was (and as her *numan* therefore was) there was little she could achieve in her old age. Once dead, however, she (or her *mini*, to be precise) was able to take hold of her son's *mini* and, developing the associative image of his final act of deceit, turn him into a human bird. Particular attention was paid to deathbed utterances among the Kakoli. Their portents were debated, for they could be 'folded talk' (*ungu iku*), that is, words with a hidden meaning (cf. A. Strathern 1975b; Merlan and Rumsey 1991: 102–09)[5] and they were revisited when explanations were sought for subsequent occurrences to which they could have had relevance. Kapiye's chronic neglect of his parents was understood by the community to reach its finale in an incident involving a bird or its eggs, and the community discounted Kapiye's own explanation about the meal of rice as an evasion of the well-known truth. The 'last words' of Kapiye's mother, which he himself did not allude to in the version of events he gave me, were pivotal in everyone else's discursive representation of his *kekelepa* episode.

Clearly, for the Kakoli, there was no doubt from the beginning that Kapiye's madness was entirely connected with his behaviour toward his mother. It is very likely that the community had expected that some affliction would be visited upon him after her death. At the core of the elaborate versions of Kapiye's period of *kekelepa* is the reasonable certainty that he spent time perched in trees (I was never able to find anyone who claimed to have seen Kapiye in the act of flying) and sometimes had to be physically restrained on the ground. While treetop sitting, in itself, may seem particularly unusual to Western readers (in advance of the conceptual complication of 'flying') it should be considered in relation to the wide geographic occurrence of treetop perching as an occasional behaviour in the general highland area. I have been told of the phenomenon occurring among groups in the Southern Highlands Province,[6] in the Simbu Province,[7] and in the Enga Province.[8] In all cases the tree-sitting was said to be related to the agency of ancestors, though precise details of the processes and reasoning involved were lacking. In addition, in the Kaugel Valley at least, the treetop area is significantly (though not at all exclusively) a domain of *kuru*. Pre-Christian practices described to me would sometimes, for example, involve the enticing of *kuru* down a vine especially arranged with one end high in a tree and the other in a bamboo container.[9]

When Kapiye was said to be sitting on branches which were normally unable to carry the weight of a human, his transformation into a bird was implicit, but so also was the agency of his mother, holding him aloft. 'Flying' or being carried from treetop to treetop was an essential discursive element in recounting the episode for both Kapiye and the rest of the community. Kapiye was always spoken of as being lifted or carried or moved by his mother. She pulled him out of houses; she flew him among the trees; she took him up Mt Giluwe; she threw him down from a great height, and so on. The reference to the stench of putrefaction (*polu*) by the young men who had tried to restrain Kapiye implied the presence of his mother's *mini* and her *popolu*, an anger which could itself become chancrous and putrefactive (reflected in the relatedness of the two terms *popolu* and *polu*).

When Kapiye was thrown down from a great height and saved before striking the ground his saviours were, according to one storyteller, other dead relatives who intervened, so to speak, on his behalf. Kapiye's own story that he was being carried by dismembered people de-emphasized his mother, but retained the sense of being tormented by the dead. The allusions here were to the complicity of *tarama* (the collectivity of ancestors who were not individually distinguishable) with his mother's *mini*, sympathetic to the *popolu* that was being transmitted by her. The involvement of *tarama* in afflictions was not unusual among the Kakoli. When living people felt wronged and developed *popolu*, it was possible for *tarama* to feel sympathy for the wronged person and act on their behalf, bringing misfortune of some kind on the transgressor, although there was no guarantee that they would intervene in this way. They did not seem to have afflicted Kapiye while his mother was still alive.

In the context of revenge and *kekelepa*, however, the madness displayed by Kapiye can be thought of as resulting from one of a number of misfortunes that his mother could have visited upon him (crop failure or bodily harm, for instance) as punishment. As she was capable of taking hold of his *mini* she chose, in this case, to torment him bodily and publicly. Kapiye's social behaviour on the ground, when he was not being 'flown' by his mother – rampaging in gardens, chasing people and so on – displayed his torment. In this respect the theatrical episodes and the eventual pig-killing and ceremonial dress were a public expression of culpability and remorse in the face of (possibly unspoken) social censure. Kapiye's attempt to avoid mention of his mistreatment of his mother during our conversation can also be interpreted as part of his *pipili*, of which humiliation and angst were significant components (demonstrating again why 'shame' is an inadequate translation).

It should be remembered that the story of Kapiye was told to me as an example of the importance of looking after one's immediate kin. Indeed Kapiye himself was complicit, both at the time and in subsequent discursive elaborations, in the construction of the episode of his punishment. Further, not only was misfortune of some kind expected by the community after the death of Kapiye's mother but anecdotally there was also a sense of inevitability about the eventual pig sacrifice. In the various versions of the story that I heard, the pig-killing was the least elaborated and most matter-of-fact element, spoken like an anti-climax: '...then

Kapiye made a pig sacrifice and returned to normal'. Psychologistic explanations might be tempting to a European commentator, such as the possibility that this was an elaborate piece of ritual theatre with 'traditional' precedents, whereby Kapiye 'acted out' his torment for as long as was deemed appropriate before the pig sacrifice. But these kinds of explanations would involve a historical and ethnographic decontextualization.

In the 1980s, when this incident occurred, the 'culture-bound syndrome' was heuristically popular in the discourse of transcultural psychiatry. I have no doubt that Kapiye would have been classified by a proponent of the concept of the culture-bound syndrome as an apt example, a short-term psychosis or hysteria through which guilt was acted out, the appropriate cultural action (pig-sacrifice) was taken, and the individual returned to normality. But such a designation would have given the episode a synchronicity and cultural typicality that it did not deserve. While Kakoli could recall, conversationally, other instances of madness as punishment of the living by the dead, the events of Kapiye's madness were singular. The mad did not always take to the treetops, madness was not always interpreted as the work of dead kin, and Kapiye's 'flying' was most unusual, according to even the eldest of my interlocutors, although tree-sitting itself was not.

For the community the singularity and drama of the episode, when recounted, was its major aspect; its uniqueness made it a cautionary tale about a fundamental principle of social life in the Kaugel Valley – the importance of fulfilling kinship obligations. Its salience was precisely that it was not a recurrent, occasional or typical 'syndrome', but a single drama of madness that could be discursively linked to other equally unique incidents of posthumous revenge to reinforce a simple point about Kakoli sociality. To refer to Kapiye's madness as a 'culture-bound' phenomenon would surely misrepresent it as individual theatre determined by formal institutional culture. Rather, the Kakoli (including Kapiye himself) actively and discursively constructed Kapiye's *kekelepa* episode in a specific historical context and in terms of an imperative of their mode of production and its contingent sociality.

Wanpis: The Victim of a *Kuruwalu*

The stories I was told about madpeople in response to my initial enquiries mostly contained a dramatic narrative, as exemplified above by the stories about Kapiye. Perhaps it was her contrasting minimal social visibility that excluded the woman I will call 'Wanpis' from these early stories. At any rate I was not told about her and it was almost two months after my arrival in the valley before I encountered her. I noticed her one day at Tambul and could easily have overlooked her, except that her singularity was noticeable in relation to a group of men clustered by the louvred window of the ADC's office trying to overhear conversation inside. A little apart, and squatting alone, the near-naked woman[10] was covered with dirt and with what looked like a scaly rash on her torso and thighs. She appeared completely withdrawn: Her morose, slightly apprehensive expression would occasionally give way to a brief giggle, as if at a humorous remark by an invisible companion. She made no response to a conventional greeting from me in her own language.

I had come to the ADC's office to obtain census data, and after a conversation with the ADC I emerged to see the woman in the same spot. I asked him who she was. 'Ah. She is one of those people you are interested in,' he remarked, but he knew little else about her. Asking around, I found her name was Wanpis and that she was from Opiapulu, an area about two miles downvalley from Tambul. She slept out in the forest and sometimes stole vegetables from people's gardens. I asked about the background to her current state and was told that she had once been married and had worked with her husband at a tea plantation near Mt Hagen. During this employment, when her menstruation had occurred she had been sent outside to sleep by her husband. Traditionally women slept in a separate house on such occasions, but there had been no women's menstruation house at the plantation and she had been obliged to sleep in the open. She had been visited by a *kuruwalu* ('feral' *kuru*), which had appeared in the form of her husband and copulated with her. After this, according to my informants, she had gone mad.

The story of the *kuruwalu* attack was reiterated several times during my later enquiries about Wanpis, with minor variations. The sexual act was described variously as a seduction (the appearance of the *kuruwalu* in the form of her husband ensuring her compliance) or as rape. Some informants who were fluent in Tokpisin substituted the term '*wailman*' or '*maselai*' for *kuruwalu*. The two Tokpisin terms are vague and general in their application and their meaning is often context-dependent. The Kakoli use of '*wailman*' resonated with the concept of *kuruwalu:* It could appear in the form of a friend or relative, lulling a person into a false sense of security before attacking. *Maselai* is a term commonly used in the sense of a dangerous 'ghost' in many parts of PNG.

During my first encounter with Wanpis one of the ADC's staff members took it upon himself to try questioning her on my behalf, but he obtained only fragmentary responses from her in Umbu Ungu. After a few moments I asked him to stop, feeling some remorse at having made her the object of attention of a small and growing crowd. Her lack of communication was her normal behaviour, I was told. She would appear around Tambul occasionally, hoping to be given food scraps, and would then disappear into the forest again. I walked away at this point: It was the only way to disperse the crowd that was by now encircling the squatting Wanpis and myself. Subsequent casual questions established that Wanpis's husband had left the area a long time ago. She was said to have borne several children since then but they had died in infancy through neglect, except for one or two that had survived and been looked after by other women (I was unable to locate these but did not discount the possibility of their existence).

A few days later I was again at Tambul and rainfall drove me with a tight crowd of people into one of the station's small tradestores. The wooden buildings contained a relatively large area for customers, since all goods (rice, tinned meat and fish, matches, soap and other small items brought by truck from Mt Hagen) were shelved on the walls behind a serving counter. Three or four of these stores stood side by side facing the currently muddy road, providing an erratic service punctuated by days or perhaps weeks of closure when stock ran out. They also functioned as shelters from the regular showers of rain, when they would quickly fill with a pressing crowd. The store in which I found myself sold hot flourballs deep-fried in oil and like many in the crowd I passed the heavy shower munching several fist-sized balls to counter the chill brought by the rain. Among the last people to straggle into the store was Wanpis, who settled just inside the doorway. She was almost naked and covered in dirt, as she had been the first time I encountered her. She looked at no one and was ignored by everyone. As the rain eased I pushed my way to the door and as I passed Wanpis I offered her a flourball as casually as I could to avoid the communication becoming a piece of theatre for the crowd (as often happened to me). She turned her head away negatively without looking at me, and I drifted off in the general exodus.

At times later in the day I noticed her squatting here and there, at the edge of a group milling at the roadside or close to the women selling sweet potatoes and green vegetables at the marketplace. She may have been hoping for food scraps, but I observed that she neither spoke nor looked directly at anyone and was completely undemonstrative. I did not see Wanpis again for a month or two, and I found that she appeared around Tambul only occasionally and was untraceable much of the time. Meanwhile I visited Goroka Hospital's psychiatric unit and discovered the 1976 file[11] whose substance I described in chapter 2. She was referred to by five different names, judged at the time to be about twenty-five years of age, medicated with chlorpromazine, and then discharged. A post-discharge letter in the file stated that she had originally been detained by police on charges of prostitution in Mt Hagen. It said her behaviour had become strange after the birth of her second child, and her husband had left her because of her irrational behaviour and promiscuity. He had taken the children with him and she had gone back to Tambul, where she had spent much of her time 'loitering' and 'consorting with men' before drifting into Mt Hagen, where she was arrested. She was described as having no schooling. The letter stated that she had been admitted as a psychiatric patient because 'her behaviours were odd, like being angry towards any-one without reasons, would cry and would fight other people without reason'. The file's version of events made no reference to stories of rape or seduction during menstruation.

I was never able to clarify the background to the arrest of Wanpis for prostitution in 1976. Nobody among my Kakoli acquaintances seemed to be interested in the official account of what had happened at Mt Hagen and Goroka and everyone seemed satisfied with the *kuruwalu* story as the explanation for her madness. The unsigned short letter in the Goroka Hospital file was the only other source of information I had. Tambul district tax census records of 1977 gave her birth date as 1951 (an approximation) and she was listed as having two children, born in 1970 and 1975 (these dates would also have been approximations). If the letter in the hospital file was correct in reporting that Wanpis's behaviour became strange after the birth of her second child, then no more than a year passed between the appearance of odd behaviour and the arrest for prostitution which led to her being sent to Goroka Hospital.

Musing on the little information I had, I allowed myself the ethnocentric speculation that post-natal depression might have been a factor. To my casual questions Kakoli women always denied that they ever felt distress or hostility toward infants immediately after childbirth, but persistence produced one or two anecdotes about women who had displayed *kondo* (sadness) or *konopu talo* ('two thoughts together', i.e. confusion) after giving birth. I also considered the possibility that Wanpis had been raped by someone (factually grounding the *kuruwalu* explanation) and that the second child had been the result. There was of course no way such speculations could be seriously investigated – and the temporal reference 'after the birth of such-and-such a child' was a common local recourse in placing past events which did not mean *immediately* after the birth.

One or two people remembered that as a child Wanpis had been clever and had gone to the (evangelical) bible school at Tambul, and could speak Tokpisin and English (this contradicted the Goroka Hospital file comment about a lack of education). One middle-aged man told me that some of her kin had built her a house after she went mad, but she had burnt it down. He added that some irresponsible men had sex with Wanpis. He offered the opinion that her husband was to blame for her madness, having sent her out into the night in the first place, and considered that her kin should have sought compensation. The story about the burnt house was later elaborated by Opiapulu people who said they had built it on instructions from the ADC of the time when Wanpis returned permanently to the valley (presumably after her release from Goroka Hospital). It was assumed that the house had caught fire because of Wanpis's incompetence.

The parents of Wanpis were dead and the community, including her natal clan, showed no particular interest in her that I could see. The next time I encountered her, again at Tambul, she was squatting in the marketplace with a vine tied around her head. This was a common response locally to *penge nokomu* (headache). I asked her if she had head pains and got no reply but my attempt at communication immediately drew attention from the crowd, as it always did when I was seen interacting with an individual whom the community regarded as mad. Some women barked at Wanpis, 'The red man is asking you something! Talk!', much to my discomfort. I told the assembly that if Wanpis had head pains I had some medicine (painkillers) that would help. This was immediately shouted at Wanpis by everyone at once. Wanpis made no response and squatted immobile, and several people commented to me that she did not want medicine. I wandered away to spare her the crowd's attention (though I had no idea how much Wanpis recognized or cared that she was being made a spectacle), and the crowd dispersed.

Making enquiries at the health centre, I learned from a nurse that Wanpis had approached them with a sore throat some weeks previously. This confirmed that Wanpis could communicate verbally to some extent, even if only in the fragmentary way that I had seen outside the ADC's office. The nurse said the rash on Wanpis's body was recent, and the doctor said it was fungoid and could be cleared up with a course of antibiotics, but nothing could be done about it unless she turned up regularly at the centre – a very unlikely prospect. Wanpis's passivity and lengthy absences from view were possibly the main contributing factors in the community's lack of interest in her. The only exceptions to this general attitude seemed to be when I was seen attempting to communicate with her.

From my observations I had developed an impression that Wanpis might be more accessible for communication than the community seemed to think, though there was no particular evidence to support the idea. I wondered whether better communication might be possible if the inevitable crowd could be avoided. Such a circumstance was virtually impossible, since I saw Wanpis mainly in Tambul, where there were always people milling around. Any other occasional encounters were on the road that led down the valley from Tambul, and there were always observers working in roadside gardens, sitting at the roadside or strolling to and from Tambul.

I managed some minor advances in communicating with Wanpis, however. One occurred as I was walking one Sunday through Tambul, which was almost deserted because Sunday was market day at Kiripia (12 kilometres down-valley) and most people spent the day there. Tambul had a small church and a man had settled himself at the roadside nearby with a few bunches of peanuts for sale, hoping to attract custom from the congregation as they left for home or for Kiripia. Three or four people sat with him talking. A few yards away Wanpis squatted; there was no one else in sight. I bought two bunches of peanuts from the vendor and then turned to Wanpis and asked her if she was hungry. She giggled without looking at me and after a pause said in a murmur that she was. It was the first time she had spoken to me. I put one of my bunches of peanuts in front of her and strolled on, while the peanut seller and his companions exclaimed, predictably, 'The red man is giving peanuts to Wanpis! The red man is talking to Wanpis!'.

On another occasion I passed Wanpis wandering on the road near Opiapulu, with two elderly women as the only witnesses. Remembering that I had been

informed that Wanpis had once spoken Tokpisin and English, I said casually in
Tokpisin 'Monin Wanpis' ('good morning Wanpis'). As though automatically,
she echoed 'Monin' (again, her voice was a murmur) and then turned her head
and, for the first time, looked at me. At that moment one of the elderly women
exclaimed loudly, 'The red man is talking to Wanpis!' and they both laughed
excitedly. Whether I could have achieved any further conversation with Wanpis had the women not been there was impossible to tell. In the event, Wanpis
looked away and wandered off. I found incidents like these frustrating. While I
remained a curiosity throughout my fieldwork period I was able to talk to most
people without attracting a crowd (at least after the first few weeks), but it was
impossible in Wanpis's case.

I developed the habit of carrying a cooked sweet potato or some dry biscuits
in my *bilum* (net bag) in case I encountered Wanpis, who was thin and looked
chronically undernourished. I usually put the food at her feet without ceremony
or comment as I passed her squatting figure, to avoid attracting attention. I noticed that she invariably ate the food after a pause. Other people occasionally
threw her scraps, but not often. She frequently wore a vine or a strip of plant
fibre tied around her head, suggesting persistent headache. Once I saw her squatting at the roadside at Opiapulu with a fibre around her head and gave her two
painkilling tablets, telling her and the group of women sitting nearby that they
were for the pain in her head. She took the tablets in her hand but made no move
to swallow them (people usually gulped pills down without water, I noticed). I
asked the other women to ensure that she consumed them and walked on. When
I returned later in the day Wanpis was gone. I was told that she had thrown the
pills away – there was some bemusement among the women at my having given
Wanpis 'medicine'.

At the end of my fieldwork I knew very little more about Wanpis than I had
at the beginning, and her condition and social relationship with the community
remained unchanged. My occasional reiterations of the material I had found in
the Goroka Hospital file were acknowledged as interesting 'stories', but had no
effect on the local consensus about the menstruation episode and did not stimu-
late any rethinking about her condition or its origin. I attempted several times
to persuade the health centre at Tambul to take the initiative in improving her
general health. The doctor seemed at one point to subscribe to the idea, suggesting that offering food might attract her regularly to the centre for long enough to
complete a medical check and even a course of antibiotics for her skin condition.
In the end, though, nothing was done and Wanpis remained a silent, shadowy
presence at the periphery of the community. To a degree the community's lack
of interest in Wanpis as a social presence reflected her inability to contribute to
Kakoli subsistence. She did not tend gardens or raise pigs or children: She had no
active social role. She did, however, have a discursive function.

In review, the Kakoli explanation for Wanpis's *kekelepa* presented her as a victim of circumstances. The framing circumstance was the lack of a menstruation house at the plantation. Her husband was also blamed for making her sleep in the open, where she became vulnerable to attack. The *kuruwalu* (or *wailman* or *maselai*) played an ambiguous role in her fate. It was a sexual predator and perhaps a cause of pregnancy (though this was not clearly articulated). It was not made clear, though, whether the *kuruwalu* had appropriated her *mini*, for her *kekelepa* was spoken of as if it were post-partum, rather than following immediately upon the *kuruwalu*'s visit. If the *kekelepa* had been short-lived, it might have borne comparison with a situation described by Marilyn Strathern (1972a) in a discussion of grievances and anger among Melpa, and sickness arising from *popokl*, the Melpa equivalent of Kakoli *popolu*. Strathern writes that Melpa women can become possessed by 'wild spirits', which can sometimes be a matter of guilt, but frequently 'is a revelation of *popokl*, the sick person complaining of a grievance rather than admitting to a sin' (ibid., 254). She cites cases in which the spirit possession revolved around grievance directed by women at their husband, and comments that the possession 'draws attention to a plight rather than inflicting damage in protest, oriented thus to excite sympathy and to avoid conflict' (ibid.).

Strathern's is an interesting point, and she compares cases of spirit-possession among men, of which fewer 'arose because of grievances against the wife' (ibid.). Yet her interpretation diverges from Melpa perceptions, and overrides their explanatory use of the 'wild spirit', inasmuch as she offers a social-psychological functionalist explanation of 'spirit-possession' madness. Nothing of the kind was offered by the Kakoli in the case of Wanpis, and to suggest that she was trying to court sympathy for a grievance against her husband's treatment of her could only be speculation. As Wanpis's *kekelepa* persisted in the long term, it seems unlikely that it was a strategy against her husband. More relevant here is the community's construction of her *kekelepa* as the outcome from an encounter with a *kuruwalu*, in contrast with the ambiguities of the accounts in the hospital file which left her 'irrational' behaviour and 'promiscuity' unexplained.

In respect of 'promiscuity', married Kakoli women in particular were quick to act on any sign of behaviour by other women that they perceived as a real or potential threat to their own marriages. Public accusations and physical attacks were often made on the basis of unfounded gossip alone, and wrongful accusations or even assaults were not uncommon. Innocent women accused of consorting with other women's husbands (regardless of whether sexual acts had actually occurred) were equally quick to demand compensation. The precipitate fury among women at any hint of individual promiscuity was an effective sanction overall. The Umbu Ungu term *wapera* was applied to females who were promiscuous or adulterous or who engaged in prostitution (professional prostitutes were also referred to using the Tokpisin term *pamuk*)[12] and such behaviour was long remembered. If Wanpis had been deliberately and visibly promiscuous around Tambul in the past, it would certainly have been remembered and added to any stories about her. No women mentioned any such behaviour, however, and anecdotally she was spoken of as a sexual target rather than as a *wapera*.

The discursive representation of Wanpis's *kekelepa*, like that of Kapiye's, drew creatively on Kakoli ontology, rather than applying a stereotype. The account of the *kuruwalu* attack did not simply reproduce a 'traditional' image. Rather it was contextualized in a situation that was contemporary and acknowledged social change, since it had taken place away from the valley on a plantation where there was no menstruation house. Some discussion of menstruation and sexuality is needed here to indicate the significance of this element of the story of Wanpis. It is common in anthropological literature to use tropes of 'pollution' in relation to women's menstrual fluids, but it is more appropriate to refer to them as 'harmful' or 'dangerous' in the context of the Kakoli lifeworld. Traditional Kakoli menstruation taboos need to be understood in relation to *mini* inheritance, gender dichotomies and sexuality in general. We have seen (chapter 3) that children inherited their *mini* through the blood of their mother and the semen of their father. The dichotomous representation of essential female and male qualities was elaborated through ideas of relative weakness (female) and strength (male), softness (f) and hardness (m), distracted ambivalence (f) and clear-headed resoluteness (m), and so on. Further, maleness was sustained by the preponderance of the male substance, *kopong*, which was particularly manifest in semen,[13] and was partly lost in sexual intercourse. This imbued the act of sex with paradoxical qualities. It was desired and enjoyable, but at the same time dangerous, particularly for men. Women were life-givers and therefore potent, but their intimate proximity was thought to be debilitating of men. The sexual act was referred to as *ulu keri* (bad activity) and was accompanied by careful attention to how frequently and where it took place, the post-coital behaviour of the couple, and the effects of the leakage or escape of sexual fluids and odours. The dangers of female *meme* (blood) were greater when women were menstruating, and they were required to be especially careful and to sequester themselves.

In the Kaugel Valley the tradition of sequestering menstruating women had become modified by the 1980s. In the clan territory where I resided a single deteriorating menstruation house remained, and was unused. The improvement of roads, the increase in strangers and fears about violence from outsiders were the main explanations offered to me for this change, though migrant experience in towns and on plantations, where there were no menstruation houses, were probably also affecting the custom. Women now confined themselves to their own sleeping rooms in the family dwelling, which contained separate rooms and two separate entrances for men and women. Nevertheless traditional restrictions on food handling and other dangerous contacts by the woman were maintained. At the plantation, though, restricted labourers' quarters would have made such strict sequestration inside the quarters difficult.

The story of Wanpis was redolent with the difficulties that arose when people attempted to deal with menstrual taboos in unaccommodating environments such as the living quarters in plantation compounds. Also, as the *kuruwalu* was said to have taken the form of her husband, there is an anomaly embedded in the story, for Kakoli men feared the blood of their menstruating wives and would not want to copulate with them, and women were held responsible for ensuring that no dangerous contact took place. At the same time there was an implication that the husband may have been blameworthy for sending Wanpis to sleep outside. The Kakoli explanation of Wanpis' madness was not simply a reiteration of a categorical 'cause'. It was a complex, imaginative narrative, in which they creatively used a familiar image, the *kuruwalu*, in the elaboration of a cautionary tale combining traditional ontology and contemporary socio-economic conditions. The story of Wanpis's menstruation and the *kuruwalu's* visit evidenced not so much a 'traditional belief' as a reflection on a contemporary problem.

Madness and Praxis

I have presented two accounts of madness here that superficially share a theme of 'possession': Kapiye was attacked or possessed by his mother and Wanpis was attacked or possessed by a *kuruwalu*. But to emphasize this commonality would commit the error of representing *kekelepa* as a condition whose causes could be categorized enumeratively as though the Kakoli subscribed to a logics of causality in relation to it. There was no set of categories such that 'spirit possession' (or some such description) could be listed as either a 'kind' of madness or one item in an enduring category of 'causes'. What we should note instead is the creativity in the Kakoli elaboration of these two examples. The variations in the story of Kapiye's madness (in which Kapiye himself was complicit) served to develop it as a drama of retribution for moral transgression. The consensual account of Wanpis's menstrual experience and subsequent madness constructed a contemporary problem using her husband and a traditional entity, the *kuruwalu*, as *dramatis personae*. It remained unaffected by my reportage of what I had found in the Goroka Hospital files.

The two cases described here illustrate, I believe, the value of contextualizing our analysis of madness among the Kakoli in the group's praxis, rather than seeking synchronic 'cultural' circumstances or explanations in the conventional sense that I have critiqued. We cannot abstract 'madness' among the Kakoli as something that can be investigated as complete in itself, and I have argued in the preceding chapter that we should not presume to be able to define it comfortably and enumerate its 'causes', either according to the Kakoli or reductively according to a psychiatric paradigm. We need to understand it as an element of a social totality involved in a process of historical change. The *kekelepa* episode of Kapiye and the seemingly permanent *kekelepa* state of the marginal Wanpis were part of the dialectical process through which the Kakoli have reflectively created and recreated themselves through time, dynamically and imaginatively internalizing social change before, during and since the colonial period.

Notes

1. Potatoes were introduced to the upper Kaugel by the Department of Primary Industries in the late colonial period as a potential cash crop, with limited success. Some Kakoli planted them in their subsistence gardens without adding the recommended fertilizers, producing only small potatoes.

2. There are differences between alpine and montane vegetation above approximately 3,300 metres and vegetation below approximately 2,500 metres. Most houses and gardens in the upper Kaugel were situated between 2,200 and 2,500 metres. In dry spells children trekked up Mt Giluwe especially to gather various normally unavailable flowering plants.

3. Graveyards were a colonial introduction. *Kiaps* (patrol officers) forced valley dwellers to bury their dead deeply and the Kakoli have adopted Christian burial procedures, though the ritual behaviour involved reflects a mixture of Christian and non-Christian beliefs.

4. The Umbu Ungu for 'heart' is *mund*. The term *mongo* glosses a number of things, depending on context, among them 'eye' and 'centre' and also 'trouble', in the sense of issues 'concentrating' at a central place. Sometimes *mundemong* is used to refer to the heart itself. The phenomenology of Kakoli emotions presented here seems similar to that of the Melpa, as discussed by Andrew Strathern (for example 1975a, 1981), and of the Ku Waru people discussed by Merlan and Rumsey (1991), although I cannot say that it is the same in every respect. My informants on this subject were ageing men of the Kepaka clan, (especially *alaye* Manenge, Tame Ralye and Silsu Karo). Younger people, while differently 'educated', still subscribed in the 1980s to the same ontology, but were less able to discuss details.

5. Merlan and Rumsey translate the equivalent Ku Waru (upper Nebilyer valley) term, *ung eke*, as 'bent speech' (1991: 102). In Umbu Ungu there is a reflexive verb *karaye* that means to bend oneself, or flex oneself (and metaphorically to bend but not be knocked over, to be 'resilient' or 'stubborn' against opposition) which can be used of people and things (trees, bamboo, etc). *Iku* is non-reflexive, and refers to bending or folding something. In the case of *ungu iku*, and the inference of concealment, I consider 'folded' to be the most appropriate translation.

6. Rev Stephen Pirina, pers. comm., 9 March 1986.

7. Dr Chris Ninkama, pers. comm., 8 March 1986.

8. Robert Head, pers. comm., 16 July 1986, and Ralph Bulmer, pers. comm., November 1986.

9. Cosmologically the space of spiritual activity or power in the lifeworld of the Kakoli and other Mbowamb people extended into the heavens (see Strauss 1990).

10. By the 1980s women had adopted European-style clothing – blouses and skirts – as everyday wear. Wanpis wore only a traditional apron.

11. Goroka Hospital Patient No. AD114-723. No file number.

12. There were no professional prostitutes in the Kaugel Valley. People made categorical distinctions between promiscuity (which might or might not involve payment by the man) and professional prostitution. Speculative judgements about women's promiscuity were often made on the basis of the kinds of clothes they wore, and young men who accompanied me into Mt Hagen occasionally to buy supplies would ask me if some of the younger European women they saw there were *wapera*, the query usually prompted by the individual woman's clothing.

13. The term *kopong* is used also for pig grease and other substances seen as promoting life, growth and health.

5

THE SOCIAL CONSTRUCTION
OF MADNESS: LOPA'S SEASON

This chapter, in which I give a third account of madness, thematically extends my previous observation that Kakoli discourses about mad individuals are also stories of themselves collectively. Just as Kakoli experiences and reflections were embedded in the discourses of the madness of Kapiye and Wanpis in the previous chapter, the story of a madman presented below reveals significant aspects of their sociality. The dialectical relationship between the *kekelepa* person and the community is particularly salient in the episode related here. The social construction of madness is explicit in the narrative of the effect of communal anticipations, reactions and recollections and the response and reaction of the individual. The question of what might have 'caused' or brought about the individual madness is less important here than the dynamics of the episode.

This episode also involves what I have called the ambivalent use of psychiatry. That is, psychiatric services were employed without a subscription to the medical-scientistic understanding of madness (as 'mental illness') which psychiatry represents. The community, or its local authorities, utilized psychiatry as a resource in the interests only of the physical restraint of the *kekelepa* person. This is not to say that such expedient strategies are adopted with fundamentally hostile intentions toward the individual. In the case of Lopa's madness a high degree of frustration and the exhaustion of social resources is evident in the appeal to psychiatric attention. But the mismatch between the medical-scientific, therapeutic project of psychiatry and the Kakoli response to what they conceive as serious madness is clear in what follows.

Lopa's 'Seasonal' Madness

The verbal list of madpeople given to me by members of my host clan, the Kepaka, at the beginning of my fieldwork included a man whom I shall call 'Lopa'. It was some time, however, before I encountered him. This was partly because he lived more than eleven kilometres away from Kepaka territory, beyond the territories of several other clans. Lopa was a member of the Yano clan, living in the head of the valley a few kilometres to the west of Tambul. In the early months of my fieldwork I was familiarizing myself with two or three other people who had been represented to me as *kekelepa,* and I was travelling between Tambul and areas down-valley from Kepaka territory. Yano and other territories beyond Tambul were as yet beyond my ambit. Another reason why I did not encounter Lopa for some time was that he was not yet mad. Lopa, said my informants, went mad once a year – at Christmas – for several weeks, and then returned to normal. For most of the year, they said, he stayed around his homestead and was a good, reliable man. At Christmas, however, he became *kekelepa,* spending his days at Tambul, singing and talking without stop, and sometimes stripping naked or decorating himself in a crazy fashion. He would steal food from the marketplace and generally make a nuisance of himself. People sometimes felt sorry for him and gave him 10 *toea* (roughly 10 cents) or a scrap of food, but he would throw them away. Then, according to the stories, he would return to normal and retire to his home again.

At the beginning of my fieldwork I asked the medical officer at Tambul if he had encountered any cases he would classify as psychiatric. The doctor, who was Papua New Guinean and from a coastal area, had been posted at Tambul for little more than a year, but he cited two or three cases, one of which was Lopa. He described Lopa's madness as seasonal, beginning about May each year and lasting for three months. For the rest of the year, he said, Lopa was not seen at Tambul: 'He is a good man, he stays at home and looks after his family'. The doctor's description of the period of madness was at variance with what I found to be a local consensus that Christmas was Lopa's *kekelepa* time. Christmas, introduced by missionaries and reinforced by colonial observance, was the only event that occurred with strict calendric regularity in the Kaugel Valley beyond the seven-day week that had also been colonially imposed. It was commonly used as a reference point temporally ('before Christmas', 'at Christmas', 'after Christmas'). Further, in PNG generally, where people traditionally did not record or celebrate calendric birthdays, 'Christmas' was used in representing (usually guessing) one's age, as for example in the Tokpisin phrase '*Mi winim faivpela-ten krismas*' ('I have achieved fifty Christmases', i.e. 'I am fifty years old'). The description of Lopa's madness as seasonal, regular and related to Christmas (or perhaps May) requires a diversion into more general matters of seasonality.

Traditionally, no clear pattern of seasons was experienced in the valley, where rain was common most of the time, though lessening during the period which Europeans would call the months of May, June and July. Frosts could occur at any time, the heaviest tending to be experienced during the period when rainfall lessened. Lunar months roughly equivalent to those of the Western calendar were recognized, but were not categorized into yearly cycles by the Kakoli. Bowers

(1968: 28) found twelve named lunar cycles to which Kakoli attached markers of a very general kind, such as 'rain', 'good weather' (applied to five moons, including three typified by constant rain), 'rain and bad weather', and 'windy', and in the case of one moon 'pigs get fat'. One event that was regularly 'celebrated' in earlier times was the ripening of mountain pandanus (*P. jiulianettii*), although the fruit did not ripen in the same moon each time. The availability of the pandanus nuts briefly eased the reliance on sweet potato, and their ripening was met with communal uninhibited merry-making. This response was also observed by colonial authorities in some other highland communities, and interpreted as a kind of collective madness (see, for example, Sinclair 1957: 35). The Kakoli desisted from this behaviour under pressure from missionaries (a 'law' was imposed, said elderly informants), without reducing their appetite for pandanus (a comment on the pandanus celebration is added at the end of this chapter).

There do not, then, appear to be 'traditional' precedents for linking individual madness to yearly cycles in the upper Kaugel Valley. Nevertheless I found the idea of seasonal madness to be subscribed to not only by the doctor at Tambul but also by others in the highlands. The officer-in-charge of the psychiatric ward at Goroka Hospital went so far as to offer me a theory about the regularity of the onset. An initial acute psychosis would occur, he said, and then the following year as the calendric time of the psychosis approached again a build-up of superstitious anxiety about the possibility of another attack would result in a reappearance of the symptoms. I found this explanation difficult to accept, given that yearly cycles were not commonly a mode of reference in the highlands for remembering past events. The officer gave a case example: a man who suffered from a seasonal psychosis and who as a result voluntarily admitted himself to the psychiatric ward to be restrained and looked after while the outburst lasted. This occurred every August, with absolute regularity, said the officer, producing the patient's file for me to check. The file, however contradicted his 'seasonal' claim, as the patient's admission rate over an eight-year period varied from four times in one year to a period of almost two years with no admission, and of a total of twelve admissions only one had been during the month of August.

In 1985, given local predictions that his madness would start at Christmas, Lopa appeared at Tambul a little prematurely. It was the second week of November and I had walked to Tambul, along with hundreds of other people from all parts of the valley, to witness a rare visit by a national politician. The Minister for Agriculture was making public-relations visits to rural areas during which he explained the government's new proposal to ban imported vegetables and the benefits this would allegedly bring to rural producers. A large crowd had accumulated in Tambul by midday, and a makeshift stage lavishly decorated with colourful flora had been prepared for the dignitaries, who had not yet arrived. I was milling with a group of men from my host clan when Lopa was pointed out to me. He was wearing bits of clothing and scraps of paper in an outrageous parody of ceremonial dress, and was dancing and singing and occasionally declaiming in a hoarse shout. I approached him and he shook my hand enthusiastically. He spoke a gibberish of Tokpisin and Umbu Ungu that neither I nor my companions could understand. Between his bursts of chatter I said I would like to talk with him sometime. The gibberish continued, but at one point his speech became clear and he stated in the same mixture of Tokpisin and Umbu Ungu that he was unable to talk with me at present because he was *kekelepa*, but we would talk when he was normal again. Surprised, and unsure that I had understood correctly, I turned to my companions, who confirmed that Lopa had indeed said what I thought. Lopa had gone back to gibberish again, and stamped off singing.

As the day went on, the milling crowd congealed in anticipation of the ministerial convoy of vehicles that was making its way up the winding road from Mt Hagen. Lopa's behaviour became parodic. At one point he held forth like a prominent man at a traditional gathering, though his 'oratory' made no sense. At another he chanted and bobbed rhythmically as though he were part of a ceremonial line of drumming men. Alternatively, mimicking common police crowd-controlling behaviour, he rushed at groups of people shouting and brandishing a long stick. The crowd viewed all this with great humour. In due course the minister and his entourage arrived and the official courtesies began. Lopa sat quietly, though prominently, at the front of the crowd while a number of speeches were made. Then a large display of local produce was placed in front of the stage, a gift 'from the community' to the minister. At this point Lopa got to his feet and stamped around the food making comedic gestures of eating, to the delight of the crowd. One or two police officers (a contingent had arrived with the minister) moved in to push him away and there was a slight altercation before Lopa stamped petulantly off shouting in Tokpisin 'Food! I'll eat!'

This scene delighted the crowd, but the ADC (a Kakoli himself) and his local officers on the stage looked uncomfortable, for Lopa's theatrics were not without significance. While the produce was being stacked in front of the minister there had been murmurings in the crowd, and one of my companions had commented cynically on the contradiction between this generous gift to the minister and the community's constant struggle to produce enough to feed itself. Lopa's antics seemed a ludic expression of the crowd's mixed feelings about the presentation. He disappeared from view after the incident.

On subsequent visits to Tambul I always encountered Lopa. Most of the time he wore rubbish as clothing and decoration and daubed himself liberally with soot and clay. He would arrive in Tambul late in the morning, having meandered from his home harassing people on the road and pestering them for money or food. He signalled his arrival in Tambul by banging the Tambul 'bell' with a stone. The bell – a metal pipe suspended outside the one-room building that served when necessary as a 'council chamber' – was a legacy of colonialism, used as an administrative clarion. Thereafter he would usually spend the day shouting, singing, and pestering people for money and food scraps, which he would invariably throw away if given. An encounter with me seemed to delight him, as it did everyone else, for he used my tolerance to comic effect. He would, for instance, pick up litter and hold it to his forehead, as though it were for example a ritual decoration or the badge of office of a *luluai* (Tokpisin: colonially appointed headman) and 'ask' (his speech was actually incomprehensible) for my opinion or approval of it. Or he would engage me in dialogue, which I did not discourage because I was interested in his utterances, no matter how nonsensical they might appear. These dialogic episodes followed a pattern which entertained onlookers: Lopa would gabble at me for a few seconds, I would turn to the audience and ask 'what is he saying?', to which they would reply with hilarity 'he's talking nonsense, that's all!'. Lopa would join in the laughter, and then continue the 'dialogue'.

In general, other people were less tolerant of personal encounters with Lopa than myself. I occasionally obliged his badgering for food by buying him a fried flourball or dry biscuit from a tradestore. He would rarely eat these, but brandished them at people and eventually threw them away. This pestering could be quite physical, and most people tended to push him away fairly roughly. On one occasion when he started pestering people in a tradestore the storekeeper vaulted the counter and wrestled him out of the store. Lopa took most of this in good part and simply continued his antics. The only occasions on which he became violent to others, I noticed, were when he was excessively mocked or teased. This usually occurred when children, who were afraid of him, would gang together and taunt him from what they thought was a safe distance. Lopa often carried a stick and would rush at the children and thrash at them angrily: On one occasion he struck a small boy on the head with enough force to knock him to the ground and stun him. His speech remained mostly nonsensical, though intelligible themes would occasionally emerge. He counted objects out loud, would pace out distances counting his strides, and would list the names of his children if I prompted him. Individual words were spoken correctly, though his phrases were mostly a jumble. His requests for food, money or cigarettes were usually effected with gestures rather than words.

People told me Lopa would return to normal 'after Christmas.' There was some concern about him damaging property, which had not occurred so far during the present outbreak of madness but had been an occasional problem in the past, when for example Lopa had reportedly burnt down a house, wrecked gardens and killed other people's pigs. The possibility and repercussions of Lopa's damaging houses and gardens and his general behaviour were of concern to the ADC, who began negotiating with the doctor and Lopa's eldest son to have Lopa sent to Goroka Hospital. Legally the ADC and doctor could have combined to commit Lopa themselves, but they were constrained by customary considerations. If anything unfortunate befell Lopa at Goroka they would be held responsible by his kin and be required to pay compensation. This initial attempt to remove him foundered when the ADC attempted to persuade Lopa's eldest son to organize the trip to Goroka (which would have reduced the role of the ADC and doctor to on-paper complicity and removed the threat of retaliation if anything had gone wrong).

The eldest son, whom I shall call Kera, told me that the first occasion of Lopa's madness had followed the birth of Lopa's third child. He did not know the date of that birth (like almost everyone in the community he did not know his own age in years), as he had been a child himself at the time. He said the madness came a little before Christmas each year and subsided after the Christmas period, disappearing slowly. His father had occasionally burned down houses and a year or two previously had burned down a bush-material church. Kera did not know the cause of the madness. I questioned him about the proposed referral of Lopa to Goroka, but he was not forthcoming on the subject, commenting only that there were problems involved.

When I saw Lopa again at Tambul after talking to Kera he was less manic and looked drawn and tired. He approached me and asked fairly lucidly and quietly for money to buy food. He wandered into the stores, was chased out of one of them and went quietly. It was now late November, and I wondered whether his mania had passed its peak. He was not daubed to his usual extent with soot or clay and wore very little mock-decoration. His speech was still garbled but he was more intelligible than previously and when he had tried all the stores and the marketplace unsuccessfully in his food search he wandered up to me and volunteered the comment that the coming of Europeans had been a good thing for PNG because they had brought useful items: he listed axes, saucepans, shovels and matches. He announced that he had five children; four boys and a girl. The first boy, Kera, was at bible school, he said. His speech then became unintelligible and when I asked bystanders if they could understand him they said he was talking nonsense as usual. After a few moments he wandered off.

A copy of the 1977 tax census roll had survived several years of neglect that had resulted in the disappearance of most former administration documents at Tambul. Checking this I found Lopa listed with an estimated birth date of 1930. The birth date of Kera was given as 1961, and that of the third child as 1968 (these dates would have been approximations). If Kera was correct in selecting the birth of the third child as a temporal marker, Lopa had first gone mad some time between 1968 and 1973, the time of the birth of the fourth child. The census roll had a column headed 'comments' which, in the case of most listed persons, was empty. Against Lopa, however, was the single word 'crazy'. A few days later I heard a rumour that Lopa had been taken to Mt Hagen in an ambulance, had somehow escaped and had rampaged around the town before being recaptured and brought back to Tambul. I went to Tambul, where I found Lopa as crazily decorated as ever, singing and dancing and behaving just as he had in the early days of his madness. The rumour proved to be more or less correct. Health centre staff had decided to take Lopa to Mt Hagen Hospital. He had been plucked from his wanderings in Tambul, bundled into the four-wheel-drive vehicle that the health centre used as an ambulance and taken to the open psychiatric ward. Lopa had absconded while his escorts were looking for the ward staff[1] and was later found decorating himself with flowers from a road berm. He was manhandled back into the ambulance, taken back to Tambul and released.

Lopa continued to plague the station thereafter, providing both amusement and annoyance with his dancing, singing, theatrics and badgering demands for food, money and cigarettes. While he was regarded as something of an institution, tension about his potential for causing damage increased when he started a fire in a disused latrine shelter. The ADC was frustrated by the failure of his attempts to prompt Lopa's family into organizing his removal to Goroka. They had rationalized their inaction to him with claims that a pig would need to be killed (i.e. sacrificed) to ensure Lopa's safety during treatment in the distant Goroka Hospital. Lopa's *kekelepa* state, they said, complicated the organization of such a killing. The ADC did not believe the pig-killing argument but diplomatically avoided saying so to Kera, with whom he continued trying to negotiate. He commented to me that community tolerance would disappear if Lopa became a general threat to people or community property.

The month of December passed and Lopa was still manic. There were times when he seemed to be returning to normal behaviour, when the singing and shouting would cease for a few days and the crazy decoration would disappear, though the nonsensical speech and daubs of soot and clay would remain. During these periods Lopa seemed tired and looked physically run down, and people would comment to me that his madness was going; but they would be proved wrong a day or two later when he would appear in full cry again. He had been crazy now for two months.

His son Kera had 'graduated' from the bible school and was now involved in the operation of one of the Tambul tradestores. When I encountered him in the store one day I asked him whether he genuinely wanted his father to be helped. Kera said he did but he seemed sceptical about the possibility of medical intervention. Attempts had been made in the past to medicate Lopa but they had not been successful. The doctor had told me that chlorpromazine had been tried during the previous time that Lopa had been mad, but injections had to be given forcibly and this had made him angry toward health centre staff. The medical unit at Tambul (comprised mostly of nurses who were themselves Kakoli) had become wary of trying to medicate Lopa because they feared retribution from him. The doctor pointed out that the regular administration of chlorpromazine (e.g. a maintenance dose twice a day) was impossible without at least the co-operation of Lopa's family to ensure he took the medication. Kera told me that his father returned home each night during his manic periods but did not sleep. This may have explained Lopa's periods of seeming exhaustion. I commented to people generally that Lopa's madness was not leaving him as predicted, though it was now 'after Christmas'. They agreed with me, but no one seemed perturbed by this academic point. The efforts of the ADC and the doctor to get Lopa to Goroka were still being frustrated by Lopa's immediate family, represented by Kera, who would never elaborate for me the 'problem' of organizing Lopa's removal to Goroka.

One day subsequently I encountered Lopa in a quieter mood. Sometimes he gave the impression that he was acting out a clown's role, responding to expectations, and could almost step out of the role for moments now and again. On this occasion he asked me where I came from and then asked me conversationally for 10 toea (a toea is a PNG cent) to buy a piece of pigmeat at the marketplace. I replied that 10t was not enough and in any case there was no pigmeat at the marketplace. 'True', he said reflectively, then added that he was a *kekelepa* man and this was why he said such things. We were strolling, just the two of use, from the quiet outskirts of Tambul toward the crowded marketplace. 'You're crazy, eh?' I asked in Umbu Ungu. 'Yes', he replied and waved his arms briefly in mimicry of his usual antics, and chuckled. I asked what caused him to go mad, and he said there was no cause. We were nearing the marketplace and I gave him 10t. He left me and shortly was striding around in the crowd shouting and waving his arms, manic again.

Weeks went by and Lopa remained mad. People were becoming tired of him and he was being treated roughly quite often now. As if in response, he became more of a nuisance, damaging trees around Tambul and smashing louvres in the administration buildings. During May, by which time Lopa had been mad for six months, he set fire to a bush-material house and was attacked by the occupier with a stick, receiving a broken wrist in the encounter. Some accounts of the incident said Lopa had set fire to the house in response to aggravation from the occupier. His wrist was treated and fitted with a plaster cast at the health centre, but within hours he had cut it off. He was now regarded as a destructive pest and his family changed their attitude to the removal of Lopa to Goroka Hospital. The problem of the pig sacrifice disappeared and they wanted him sent for treatment, but were unable to organize the trip.

Immediate efforts by the ADC and doctor to organize the 'referral' broke down through bureaucratic tangles with the Mt Hagen and Goroka hospitals and transport problems. Meanwhile Lopa's destructiveness lessened and the tension over his behaviour decreased. In July he stopped appearing in Tambul. I found Kera in his store: He said Lopa's madness had gone and he was now at his home. It was eight and a half months since the onset of his manic behaviour. Kera was agreeable to my talking to Lopa at his house, which I found after a long walk and guidance from Yano clanspeople. Lopa's wife and children were there, but not Lopa. He had gone off to sleep in the forest, they said; he always disappeared into the forest for a short time and then returned home. Lopa's wife displayed a resigned attitude to his madness; it had started a long time ago, she said, without apparent reason and simply came and went all the time. No one had ever been able to do anything about it. His second-eldest son, who was present at this conversation, said they had on occasions tied him up in the house, but he would eventually get free or make such a noise that they freed him anyway to get some peace.

The son said that when he was a child his father's outbursts of madness had been relatively short with long periods of sanity between them, but gradually they had become longer in duration and closer together. 'Medicine' had been tried on Lopa but had worn off quickly. At one stage they had been given a supply of medicine (probably chlorpromazine) to administer to Lopa but he had refused to take it. They were concerned about Lopa's potential for disruption in the community at large and the possibility of claims for compensation. He had become more prone to violence and destructiveness in recent times, they said. I mentioned the proposals by Tambul authorities to send Lopa to Goroka for treatment, but the younger brothers clearly deferred to Kera in the affair and they told me I would have to speak to him to find out what was happening. I arranged to return to their house in a few days' time when, they assured me, Lopa would be back from the forest and I would be able to talk to him.

Several days later I set out from Kepaka territory for Lopa's home, only to be confronted when I reached Tambul marketplace with the spectacle of Lopa covered in soot, clay and several varieties of flora in the act of disrobing to the laughter and shrieks of a large crowd. I found Kera in his store, and he informed me that Lopa had gone mad because of some 'medicine' given to him by the health centre. Checking at the health centre I learned that the doctor had tried a strategy of starting Lopa on a maintenance dosage of chlorpromazine while he was sane, with the idea that the dose could be increased if Lopa showed signs that the mania was coming on. After two days of the medication (administered by his sons), Lopa had gone mad. Kera was understandably blaming the medication, but whether the pills had actually been administered correctly (or whether Lopa had swallowed them) was impossible to know. In a few days Lopa's antics had taken a serious turn and anger was once again generating around Tambul. He had smashed all the glass louvres lining the upper side walls of the small Catholic church on the outskirts of the station and burned down a nearby latrine-shed. The doctor told me he had been keen to have Lopa apprehended and given a large dose of chlorpromazine by injection, to start a closely monitored régime of the drug. However, he could not persuade staff to help in the venture because they all feared retribution and damage to the health centre when the initial effects of the medication wore off. Attempts were also being made once again to organize the removal of Lopa to Goroka.

The possibility of Lopa provoking violent retribution, demands for compensation or more serious interclan problems seemed stronger than ever now. The complex dynamics of the interaction between the ADC, the doctor and the re-calcitrant Kera appeared no closer to resolution than they had ever been. Even if the doctor were able to start the régime of chlorpromazine, its continuation as a twice-daily dosage seemed unlikely in view of the family's negative attitude to medication and Lopa's probable reluctance to take it. The removal of Lopa to Goroka seemed unlikely. If tranquillizing drugs were the only way to keep Lopa's behaviour from ending in major trouble or personal tragedy the doctor would need to overcome the problem of regular administration. The doctor, like most in PNG, had no training in psychiatry, and had never administered any psychi-atric drugs apart from anti-epileptics and chlorpromazine, which was part of standard medical supplies in regional centres. He learned during a conversation with myself on the subject that there were alternative tranquillizers to chlor-promazine, including the phenothiazine fluphenazine decanoate, which could be administered on a monthly, rather than twice-daily basis. Subsequently a small supply of the latter drug was found to be held at Mt Hagen Hospital, and two ampoules (enough for two single monthly injections) were procured. However they were not immediately administered, and when my fieldwork period ended in late 1986 and I left the valley, Lopa was still *kekelepa*.

The Social Construction of Lopa's Madness

Local anecdotes and a brief encounter might have persuaded a casual observer that Lopa was given to a seasonal madness but as we have seen, under prolonged observation Lopa's madness did not lend itself to such simple categorization. On the evidence of the old census roll and of Kera's childhood recollections Lopa had probably been periodically going mad for more than a decade before I arrived in the Kaugel Valley. We can make no firm statements, though, about the duration of the periods or the regularity of their occurrence. The account given by Lopa's wife of the periodicity of the madness was at odds with the popular representa-tion that oriented it to 'Christmas'. Given that Christmas was the only annual calendric reference point available to the Kakoli, it is not surprising that intermit-tent madness lasting more than a few weeks was discursively represented to me by them as always occurring in relation to the Christmas period. Whether it lasted for a number of days or several moons, or the ten months which I witnessed, at each occurrence made little difference to them in retrospect.

If we were to capitulate to a clinical perspective, interrogating Lopa's *kekelepa* behaviour as a 'psychotic' episode, we would be obliged to note that the behav-ioural period which I witnessed was internally inconsistent and that its duration may have been affected one way and another by the efforts of the ADC and the doctor to restrain Lopa. For there were times when he appeared to be returning to normal, only to relapse into madness, and the interventions of the ADC and the doctor occasionally served to exacerbate the situation as he reacted to their attempts at medication. The interventions moreover raise a further diagnostic complication, as they represent a recent influence on the nature and course of Lopa's madness, where in earlier times more traditional forms of restraint had been used. These clinical qualifications, of course, infer the shortcomings of the transcultural psychiatric paradigm insofar as they oblige us to consider the dynamic social context of Lopa's madness. For, as we can reasonably conclude, his *kekelepa* periods were not a cyclically regular 'seasonal' outbreak of prescribed duration. Nor were they framed by a synchronic 'cultural context' in the sense that cross-cultural or transcultural psychiatry attempts to acknowledge even as it moves beyond assumptions that local culture is a clearly demarcated entity with a specific set of values and practices (see chapter 2).

Some of Lopa's behaviour can be interpreted as parodic responses to particular circumstances, such as the 'official' visit by the politician where he declaimed like a big-man and mimicked crowd-control behaviour, and my own presence where some of our interaction parodied colonial engagements between local spokespeople and patrol officers. But it would be unwise to romanticize these instances, or the moment of ludic honesty involving the food display for the politician, as if they were oracular in the sense suggested of madpeople by Foucault (1982, 2006) or Laing (1973). To do so would be to abstract them from the dynamic social process that included Lopa's attack on children, smashing of louvres, burning of buildings, badgering demands on passers-by for food and money and periods of lucidity. Dialectically, we must also consider the actions of others, including the anticipation of Lopa's *kekelepa* period and the typical behaviour ascribed to him discursively by the community in advance of his appearance in Tambul. In this respect, his actual behaviour – which included most of what I was told would occur, but significantly a number of things that could not be predicted – has to be understood, not as a simple repetition of past madness, but as a singular episode in the praxis of the community.

Substantively the anticipated period of Lopa's *kekelepa* behaviour which I experienced was necessarily different from previous periods, when its course might have been affected by, for example, the presence or absence of colonial officials or medical staff, the concern or indifference of these, the availability of medication, the kind of damage done by Lopa, his (or his clan's) relation to those people he antagonized. This time my own presence became part of the dynamic of Lopa's madness, as did the actions of the ADC and the doctor, and their appropriation of psychiatry's tools of restraint (medication, the potential of incarceration in Goroka's secure ward), to which Lopa reacted variously. In recent episodes, controlling the seriously disruptive aspects of Lopa's behaviour had come to involve the attempted use of tranquillizing medication, or the secure facility at Goroka Hospital. Yet this was an ambivalent use of psychiatry, an appropriation of its practice without a care for its theory or the 'culture' – the particular history of social, political, economic and ideational developments – in which it should be contextualized. Psychiatric restraint was an alternative to binding Lopa, or to locking him in a house, and perhaps to having to give compensation for damage he might cause. No one among the Kakoli offered an explanation, or a diagnosis, in 'cultural' terms for his madness.

A Comment on 'Pandanus Madness'

A syndrome with a regular annual onset and disappearance is discursively attractive, as evidenced by its uncritical acceptance without any firm evidence by some members of the medical profession in PNG. Imaginative precedents for the idea of 'seasonal' madness have been established from colonial times, when 'amok' behaviour, often glossed as a culture-bound syndrome (see several examples categorically discussed in Burton-Bradley 1973: 63–81) and collective mushroom and pandanus madness became part of the exotic zoologization of Melanesians. Indeed, 'mushroom madness' and 'pandanus madness' became the classic given examples of 'collective hysteria' during the late colonial period. Paradoxically, considering the casual familiarity of ethnographers, medical personnel and lay Europeans with such 'madness', witness-accounts of any scholastic value are actually rare. By the 1950s – the Golden Age of highland contact ethnography – missionary activity and the post-war build-up of colonial control was already curbing the less 'civilized' activities of Melanesians and phenomena such as mushroom/ pandanus madness had either stopped or suffered a drastic reduction in participant numbers.

One of the best accounts available is supplied by Marie Reay, from fieldwork among the Kuma of the Wahgi Valley in the mid 1950s. An outbreak of 'mushroom madness' (to which Reay sensibly attaches quotation marks) in 1954 is briefly referred to in her general ethnography (1959: 188–90) and more detailed accounts are given in subsequent publications (1960, 1965, 1977). The madness involved self-decoration, dancing, brandishing of weapons and threatening behaviour which, Reay noted, seemed to be easily thwarted. Reay, who saw the Kuma as distinguished by 'strain, tension and disharmony' (1959: v), opined that mushroom madness offered 'an unparalleled opportunity for social catharsis' (1960: 139). Hers was a pressure-valve theory fairly typical of social anthropology of the time. The Kuma attributed their behaviour to the eating of a mushroom-like fungus. This was eaten at any time but at a certain time of year it was said to cause temporary madness in some people (Reay 1959: 188). Reay attempted to send a sample to Port Moresby for testing but it was found to be 'unsuitable for analysis' (ibid., 190n). Reay did not say whether she tried eating the fungus herself. Only thirty people out of a group comprising 313 were affected in 1954, and apparently there had been more in previous years (ibid., 189).

On a later occasion Reay and the psychiatrist Burton-Bradley experimentally fed alleged 'insanity' mushrooms to two Papua New Guineans 'with negative results' (Burton-Bradley 1973a: 67). In her last publication on the subject Reay wrote that her earlier reference to the mushrooms as 'hallucinogenic' was incorrect (1977: 57) and attempted to critique Kuma attribution of 'neuro-toxic' (ibid., 65) effects to them. She now saw the madness as having been ritualistic, and laid more emphasis on its socially cathartic function and its relationship to 'the difference between and the separation of the sexes' (ibid., 66). A more recent attempt to retrospectively identify the species of the mushrooms consumed by the Kuma and determine whether they were hallucinogenic or not has been inconclusive (Treu and Adamson 2006).

In the Kaugel Valley, according to older residents, large numbers of people used to become *longlong* (notably, they used the Tokpisin term) whenever the mountain pandanus nuts were ready to be harvested. This phenomenon had stopped by the time Nancy Bowers began fieldwork in the valley in 1961. She told me, however, of noting one small unexplained outbreak of collective madness in February through March of 1963 involving mostly adult males, in which the behaviour was similar to that described by Reay.[2] While older members of the community told me their *longlong* episodes had coincided with the pandanus harvesting season, none that I spoke to attributed it directly to eating pandanus. They spoke of the *longlong* as simply occurring when they went to harvest the nuts. While pandanus harvesting and collection continued to be a major activity when I did fieldwork in the 1980s there were no signs of madness connected with it during the early 1986 harvesting that I witnessed. I consumed large quantities, both raw and cooked, without experiencing any ataxia. The 'madness' (if, indeed, it can be so called) was spoken of fairly abashedly by elderly Kakoli informants as an aspect of pre-Christian behaviour that was discontinued under the influence of missions. Missionary bans on the behaviour were referred to as 'laws', and Kakoli were told they were 'sinful' (for which missionaries used the Umbu Ungu term *keri*: 'bad') and should be forgotten.

It is likely that a *post hoc* fallacy about the connection between pandanus and madness developed in colonial discourse through a primitivist bias in European observers of flamboyant Melanesian behaviour. The Sinclair Report of 1957 (see chapters 1 and 2) mentions, for example, reports of 'mental changes occurring after eating a certain variety of pandanus nut' (1957: 35) and goes on to record that:

> [d]uring the pandanus season (September to January), natives may gorge themselves on the nut and a small percentage of them become excited, restless, threatening, or actually dangerous. An observer who has lived in New Guinea for a number of years states that the condition lasts about 12 hours. He records that natives dance violently, become uninhibited and may attempt to attack one another. He recalls in the Highlands several men falling from rope bridges and drowning when in a confused ataxic

state. It is probably that there is a strong hysterical factor in the production of excitement by the pandanus nut. One observer reports that the native, once stimulated, runs about telling his friends that he has 'karuka [Tokpisin: pandanus] madness', and often his friends – normal until this time – become excited and join in the uninhibited behaviour. (ibid.)

There is much to suggest, in fact, that both mushroom and pandanus madness owe nothing to toxicity. Burton-Bradley, in respect of experimenting on highlanders with alleged 'insanity' mushrooms, commented, 'The relationship of this ... pseudo-amok to the mushroom is doubtful' and observed that some people 'indulged in the behaviour without having eaten the mushrooms at all' (Burton-Bradley 1973a: 67). Ralph Bulmer told me that 'pandanus madness' among the Kyaka Enga was discontinued in the early 1950s after a missionary nurse declared it nonsensical and administered sedative injections to some participants.[3] The disappearance of the madness in the Kaugel Valley, while the extensive seasonal consumption of pandanus continued, also indicates that toxicity was not really a factor. Bowers mentioned to me the attribution of the madness to the consumption of *unripe* pandanus, and that temporary madness could also be connected with the consumption of unripe taro.[4] But the expertise of the Kaugel people with respect to pandanus, at least, seemed to belie such an explanation. The ripening of the fruit was observed with growing excitement, but knowledgeable judgements were made about when it was ready to be harvested, and it seemed to me to be extremely unlikely that large amounts of unripe pandanus would be eaten by accident. Coupling this with the point that older Kakoli informants did not attribute their former 'pandanus madness' directly to eating the fruit (implying that they would not have deliberately eaten the unripe nuts), it seems the pandanus itself, in any state, was not a toxic cause. Former occasional consumption of raw taro (which is, actually, toxic if uncooked) was spoken of reluctantly, as an example of pre-Christian 'bad' behaviour. It was difficult for me to gauge whether the act of eating the taro was a precondition or a contingent of any state of madness.[5]

The importance of pandanus in Kaugel Valley subsistence in former times was briefly mooted by Bowers in her discussion of the postulated time at which the sweet potato was introduced to the Kaugel Valley and the effect of cultivation activity on the ecological environment (1968). She suggested that if lower montane forest had once extended down to reach what was, by the 1960s, the garden-settlement area, the wild pandanus would have been approximately twice as plentiful as it was when she began fieldwork (ibid., 239). She commented further, 'The scheduling of subsistence activities must have been determined by the ripening date of forest products, especially pandanus, to a greater extent than it is at present' (ibid., 243–44). The importance of pandanus was also reflected in the special 'language' of substitute words used by people in former times when engaged in harvesting it. In part this language, known as *amu ungu* ('pandanus talk'), served as a code to conceal, as much as possible, the intent of the harvesters from the forest *kuru* who might try to thwart the enterprise. Similar codes referred to in the literature as 'pandanus language' were used by the nearby Kewa (Franklin 1972) and other highland groups (see Pawley 1992). Before the introduction into the upper Kaugel Valley of commodities such as rice and tinned fish the huge seasonal pandanus harvest would have relieved the pressure on the sweet potato, giving them a chance to regenerate.

There is a possibility that descriptions of pandanus madness as 'amok' or 'pseudo-amok' (Burton-Bradley 1973: 67) or as 'hysteria' could have been ethnocentric exaggeration. The behaviour reported by colonial observers such as brandishing of weapons, for example, was not necessarily cause for general alarm. So-called 'weapons' (axes, machetes, spears, bows and arrows) are much more to hand in rural PNG than in European societies, and carried as a matter of course by most males. During fieldwork I frequently saw axes and machetes in particular being brandished in anger and mock-anger, and gradually became almost as unconcerned as my Kakoli acquaintances. The 'pandanus madness' could have been a celebration of the seasonal arrival of this luxury food and an acknowledgement of its important effect on valley subsistence. The merit of including it (along with 'mushroom madness') in psychiatric discourse is questionable, at least.

Notes

1. The psychiatric ward at Mt Hagen Hospital was empty of both staff and patients most of the time. I was never able to contact the officer-in-charge despite repeated attempts during my fieldwork period.
2. N. Bowers, pers. comm., 1987.
3. R. Bulmer, pers. comm., 1988.
4. N. Bowers, pers. comm., 1987.
5. Ralph Bulmer (pers. comm., 1988) told me that during a visit to the Wahgi Valley Kuma some years earlier, he had witnessed people biting into raw taro while already in an excited condition.

6

The Social Construction of Madness: The Mad Giant

In this chapter a story of madness is again told which reveals as much about Kakoli sociality as it does of the person whose madness they narrated. And once again the dialectical relationship between the *kekelepa* person and the community is particularly salient. A contrast with the account in the previous chapter is that where no 'cause' was mooted by the community for the madness of Lopa, the question of causation is an interesting part of the story below. Further, the community's portrayal of an individual as violently and dangerously mad was not matched by his actual behaviour, raising questions about why the community found it necessary to perpetuate such an apprehension of him. The ambivalent use of psychiatry was also a part of this episode. In this case a degree of deliberate manipulation of the psychiatric process to personal ends is implied. This is not to say that such expedient strategies were always adopted with fundamentally hostile intentions toward the individual: We saw as much in the case of Lopa. But the case of the mad giant involved different communal sentiments.

Hari: The Mad Giant

The example of Kapiye in chapter 4 indicated the Kakoli preoccupation with themes of social obligation and sharing and the censure and punishment, sometimes direct and sometimes subtle, visited upon those who default on such obligations. We have also seen that the Kakoli were relatively accepting of crazy behaviour, both chronic and temporary, so long as no significant personal or material damage was caused. The harsh environmental conditions in which the Kakoli people lived, their strong ethos of communality and social obligation and the unsystematic and flexible nature of Kakoli notions of, and attitudes toward, madness, are vital considerations in interpreting the narrative which follows here. It concerns the interaction between the community and Hari (not his real name), who was considered mad, and my own attempted, naïve and largely ineffectual, intervention.

I did not encounter Hari in person until several weeks after my arrival in the Kaugel Valley, but I knew of his existence from the second day of my fieldwork, for he was included in the initial anecdotal catalogue of *kekelepa* people provided to me by Kakoli interlocutors. He was, they said, a giant with the strength of several men. He could carry huge trees on his shoulder and worked harder and faster than anyone else; his strength and energy were intimidating. He ran everywhere, and when he was not running he walked very fast. He thought nothing of walking to Mt Hagen (more than 65 kilometres by winding road from his home) and would often run there. On one occasion he had walked home from the coastal town of Lae – more than 480 mountainous kilometres by road – in 'one night' (i.e. two days). Local consensus had it that Hari had become mad after a trip to Bougainville, where he had worked at the copper mine. It was speculated that he had been poisoned by the Europeans with whom he had consorted, perhaps because of jealousy over liaisons with white women. Alternatively some people suggested, as Hari boasted that he had smoked marijuana in Bougainville, that overindulgence may have been his downfall. Apart from a handful of young people and migrants, the people of the valley had no experience of cannabis at the time of my fieldwork,[1] but knew it was a drug used by some Europeans, and it was popularly believed that rural Melanesians risked terrible afflictions if they used it.

When I asked for examples of his madness I was told of his constant running and fast walking (running for any distance was abnormal behaviour among the valley people unless there was an emergency of some kind; the usual gait was a stroll). He also chopped down trees for no reason and killed other people's pigs. He talked strangely much of the time. Recently the police had taken him away to Goroka but he had escaped from their custody and returned to his home. These characteristics, then, formed the initial persona of Hari the madman presented to me by the community: a combination of superhuman physical attributes and strange, potentially violent, behaviour.

A few days later when I visited the doctor at Tambul for the first time he told me that very few cases of madness came to his attention, then asked me if I had heard of a man called Hari. I said I had heard some stories about him but had yet to meet him. Hari was *longlong* (he used the Tokpisin term) and prone to violence, said the doctor, and had recently been sent to Goroka under police escort for psychiatric attention. He had broken out of the ward – 'he is a very strong man' – and made his way back to the Kaugel Valley, and now he was avoiding agents of the law. There were problems in his marriage, the doctor told me; in fact his wife had approached the Tambul health centre to ask for his removal because of his unpredictable behaviour and her fear of violence toward herself and their children. The doctor warned me that if I met Hari I should not mention his being *longlong,* because this made him angry and violent. Hari would want to see me, because I was a whiteman, he added.

I made a number of attempts to contact Hari, but without success. He seemed constantly on the move and, unusually for the Kakoli, no one ever seemed to know where he had gone or why. Meanwhile the escape from Goroka Hospital was being elaborated in community gossip as a Samsonic example of Hari's strength. I enquired after his wife and was told that she lived by the Catholic Mission at Kiripia, but on going there I was told she was in Mt Hagen with the children. After much fruitless enquiry, I was told one day that Hari was expected to attend the Sunday morning service at Kiripia mission and that I should be there if I wanted to see him. The Kiripia Sunday service was a regular major social occasion. It was difficult to gauge how much this was due to the service itself, with its Catholic ritual, bible readings and songs in Umbu Ungu, all of which had a heightened appeal in a society whose traditional ceremonial activities and rituals were inexorably disappearing. But its social significance was also due to Sunday being market day at Kiripia, and the service served as a preliminary gathering point from which the crowd would slowly drift to the nearby ceremonial dance-ground, which became a marketplace for the day.

Inside the church the women and small children sat on the left and the men on the right, a spatial ordering of gender observed also in family houses in the valley. Entering after the service had begun, I stood just inside the door. It was common knowledge that I wanted to see Hari, and if I had not immediately been able to pick him out among the men, the host of signals from the congregation would have left me in no doubt about which was he. As I had expected, Hari was not a giant, but he was six feet tall and towered over everyone in the congregation. He was possibly the tallest man in the valley. His height was supplemented by a stocky, muscular build, so that although he was my own height he seemed massive when viewed among his own community. His physical singularity was enhanced by his skin colour, which was lighter than normal and had a reddish tinge. Hari was occasionally referred to locally as *kondole,* the colour term in Umbu Ungu referring to a spectrum from pink to dark red – the same term was used for Europeans.

I noticed in the church that Hari had some idiosyncrasies which distinguished him from the males around him and which seemed oddly European. Unlike the other men, Hari sat on the pew with his legs crossed and his hands clasped around his leading knee. And unlike the other men he seemed uninterested in the activity and speeches at the altar, and did not sing. Instead he stared around the church above the heads of the congregation. His blink-rate was very fast. He caught my eye during the service and gave me a nod. As the service finished he strode up to me and greeted me in English: 'Ah. Yes. Hello. We're going outside, are we?'

Outside the church we stood surrounded by a huge audience. I was to remain a curiosity throughout my stay in the valley, but at this early point the interest was especially great, and the meeting between myself and the madman Hari was clearly momentous. Unfortunately it made communication between us difficult (from my point of view, at least) and our conversation was limited to my introducing myself to Hari and saying, for want of anything better, that I understood he had been to Bougainville at some time. He told me yes, he had worked at Bougainville, and knew many Americans. I was to learn that Hari referred to all white people as 'Americans', and was keen for his familiarity with them to be known. I noticed during our short conversation that Hari was slightly dysphasic, mostly with pronouns, though his speech made general sense. He spoke a mixture of English and Tokpisin but not in the usual idiomatically fluid manner of multilingual Papua New Guineans: Hari lurched unpredictably between the two from phrase to phrase or from word to word. His gestures were flamboyant and he would occasionally execute a shuffling dance as he spoke. My impression from this first meeting was of a highly energetic and nervous man, eccentric perhaps but not (to *my* sensibilities) a madman. Our audience, I noticed, laughed at Hari behind his back, but not to his face. The two of us walked to the marketplace but our huge entourage was so obtrusive that I took my leave after telling him I would visit him at his home in a few days' time.

Before that visit, I learned from the ADC at Tambul that Hari was attempting to prepare a case to gain custody of his two children (a boy and a girl) from his estranged wife. The ADC told me that Hari had on one occasion gone to Mt Hagen, where his wife was staying, and taken his son away, but the police had returned the son to the mother. Consequently Hari had started a legal battle to get the two children. He was trying to strengthen his case by arguing that he was not only married according to 'custom' (i.e. by the payment of a brideprice, etc.) but had also had the marriage legitimized by the Catholic Church: Thus his wife, he claimed, could not simply leave him using customary procedure but would have to obtain the Church's consent or go through formal legal channels to end the marriage and take the children. The recently appointed ADC, being not only Kakoli but also related to Hari's wife, was in a potentially embarrassing position since he was anxious not to show clan or personal bias in his administration of the area. He had been instrumental in directing Hari's efforts into Church and legal hands in Mt Hagen, to maintain his own neutrality in the affair. I began to wonder whether Hari's admission to Goroka psychiatric ward a few weeks previously had been something other than the necessary incarceration of a violent madman. It could possibly have been partly orchestrated by his wife and her family with the unwitting co-operation of medical authorities at Tambul and Mt Hagen, building on the local perception of Hari as mad, though her initial decision to leave him could well have been because of his mad behaviour.

A few moments after leaving the ADC's office, I encountered Hari for the second time. He had run to Tambul that morning from his home (a distance of sixteen kilometres), he told me. He was relaxed, cheerful and showed practically none of the 'nervous' behaviour I had noticed at the church service. There were no other people within hearing distance and we were able for the moment to talk more easily than previously. He mentioned that he had once been to Port Moresby, and said that in Bougainville he had worked as a security guard for a catering firm. He had learned English at a community school as a child. I asked why he had left the job at Bougainville; he said in an offhand manner that he had just decided to come home. He was again idiolectically mixing English and Tokpisin; the English words and phrases were sophisticated (too much so to have been learned at community school), his dysphasia with pronouns was again evident. I asked what he was doing at Tambul that day, and he said he had come to see the police officer about a matter to do with his wife and children, but he looked uneasy so I changed the subject and asked why he ran everywhere. 'To keep fit,' he said. I commented that it was very unusual in the Kaugel Valley, no one else seemed at all interested in distance running or the idea of physical fitness. Hari agreed; he had started running, he said, after returning from Bougainville.

By this time our conversation was attracting a small crowd. Hari was beginning to display the nervous behaviour I had seen at the church; his blink-rate had increased, his gestures became more flamboyant, and his dysphasia had become more marked. I told Hari that I, too, was a keen runner and suggested that we run back from Tambul together – my house was at Lakope, about six and a half kilometres down-valley and on the way to his home. My own tendency to run everywhere was a point of bemusement to the Kakoli. It was partly a matter of expedience for me; the valley was rain-soaked for a large part of the year and I was attempting to cover a large ethnographic area. Walking from place to place was time-consuming and often meant a drenching. Running was also a keep-fit gesture in a project that involved much sedentary conversation and rain-bound inactivity. For a habitual sea-level jogger there was also the challenge of running in hilly territory at an altitude in excess of 2,000 metres. Occasionally young men would begin to jog beside me when I was running, but would stop after a few hundred metres. Without a preoccupation with fitness, and seeing no obvious reason for hurrying over a great distance, it was probably hard to see the point of forcing oneself to run when feeling the strain. I could understand that the Kakoli regarded running for longish distances without an apparent practical reason as abnormal (my own idiosyncrasies were tolerated without comment to my face, but I occasionally wondered how the community viewed me, in this respect).

As we jogged along the mud road Hari commented that he found the slow pace at which everyone in the valley habitually moved frustrating, along with the endless small-talk and story telling. This, he said, was a reason why he preferred his own company, and walked at a fast pace 'to leave them behind'. When we got to Lakope I stopped, tired from the hills and still unused to the altitude. Hari simply increased his pace and disappeared down the road at a fast trot. After these two encounters with Hari I began to suspect (with no hard evidence, of course) that he might have a disorder with an organic cause and his 'madness' may have been exacerbated socially. Everyone I spoke to insisted that he had been normal before his Bougainville trip, from which as far as I could understand (local representations of time being calendrically inexact) he had returned about five years ago. Hari's dysphasia raised a possibility that he had suffered neurological damage and I mused whether his occasional reported violence (of which I had seen no sign) was epileptic and a consequence of some malady suffered at Bougainville, and to what extent it might be socially induced.

Two days later, a time fixed during our run from Tambul, I visited Hari at his home. I had not yet told him the nature of my research and the agreed purpose of the visit was for Hari to show me around the area: he was also plainly keen to be visited by a European. He met me at the roadside and led me down a pathway, asking me for the first time why I had come to the valley and why I was living in a 'traditional', rather than a *kapa* (Tokpisin: metal-roofed) house. I explained that I was an anthropologist and that I was interested in the behaviour and treatment of madpeople in his society, elaborating a little on psychosis and psychiatry in the West. Hari nodded during the explanation and made verbal gestures of understanding, but did not comment. We passed his house, which from the outside seemed typical of the area and stood beside a flourishing vegetable garden – I recalled earlier comments people had made about Hari's capacity for hard work and the speed with which he could prepare a garden. He showed me a nearby ceremonial dance-ground and some views up and down the valley from vantage points. He was affable and speaking in his distinctive mixture of English and Tokpisin which, though clumsy, was coherent. His dysphasia was almost exclusive to pronouns, confusing 'I' with 'you', 'us' with 'them' and so on. As we chatted, I asked about his Bougainville trip. Hari said he was there from 1973 to 1979 and then had gone to Port Moresby for a few months, where he also worked as a security guard. I said I had heard rumours that he was 'poisoned' by Europeans in Bougainville. This amused him: He said he had smoked marijuana with 'Americans' while he was there but had taken no other drugs. People in the valley did not understand 'Americans', he said, and their predilection for stories and story-telling led to misconceptions.

I said I had been told extraordinary tales about his running abilities, and recounted some of them. Hari said he had never run to Mt Hagen, but had walked there a few times, stopping overnight on the way.[2] He laughed outright at the tale that he had walked back from Lae overnight. People made up stories about him, he said, because they never asked him direct questions. 'They don't talk to me' (his pronoun dysphasia made this last statement equivocal: I wondered to what degree Hari talked to 'them'). He began to develop the theme that I had heard during our run previously. Kakoli spent too much time sitting around talking and telling stories; this led to all sorts of trouble and wrong thinking. He had no time for that sort of thing. People should keep themselves occupied and work hard. He drifted on to the subject of running. This was one of the things that set him apart from other people, he said; he wanted to keep fit and active. I did not challenge his unflattering portrayal of his fellow Kakoli as lazy and inactive, but commented that people seemed a little afraid of him. Yes, agreed Hari, that was true. Disregarding the advice of the doctor, I said that people told me he was *longlong* (I used the Tokpisin term as we were speaking in a mixture of Tokpisin and English). Hari laughed and said he was not *longlong*, but people said he was because he was different and they did not understand him.

When it was time for me to return to Lakope, Hari asked if I would be running home, and said he would run with me. We set off at a fast pace; Hari seemed exhilarated and whooped and waved his arms at anyone we passed. We raced up a hill and I had to stop at the top, apologizing that I needed to recover my breath. Hari said he was tired too, but it was obviously a polite gesture, as he was breathing easily. After a few moments, he said he was going back to his home, and loped back the way we had come. My previous intuition that Hari had suffered an organic malady and some neurological damage was stronger after this encounter.

Hari's 'outsider' status became increasingly noticeable on my subsequent encounters with him. He was always alone, and never seemed to be engaged in conversation with anyone (except myself) for more than a few moments. On market days (Saturday at Tambul, Sunday at Kiripia) Hari's huge solitary figure contrasted with the group behaviour of the majority. Men moved in clusters, or in pairs hand in hand (a mark of kinship, particularly among younger men), and women would sit in small groups, but Hari would hover, staring into the middle distance in his characteristic way, always alone. I noted that the gaze of the crowd would invariably result in his making fluttering hand movements and the shuffling dance-like movements with his feet. I was far from satisfied with what I knew of Hari and his history and I was beginning to see him as, somehow, a victim rather than irrevocably mad. I began making more detailed enquiries and my conversations with Hari contained increasingly pointed questions.

I wanted to know more about the Bougainville and Port Moresby episodes, but Hari would always reply to my probing in vague terms. Something was being concealed, I sensed. I also wanted to know more about the psychiatric referral to Goroka Hospital. I had not yet asked Hari directly about this, but the subject arose one day when I met him on the road. He was running but stopped to greet me (I was constantly awed by the speed, effortlessness and distance of Hari's running – there was never a sign of perspiration and his breathing was always perfectly normal). I mentioned that I would shortly be making a brief trip to Goroka to do some research at the hospital. Hari said conversationally that he had been there recently. He had gone with a policeman friend to have a look around the town, had slept a couple of nights at the hospital and then come back. I asked why he had stayed at the hospital; had he been sick? No, he replied casually, it had just been somewhere to sleep. I knew from the ADC and medical officer at Tambul that Hari's visit to Goroka was an involuntary psychiatric referral. I had no idea whether Hari's version of the incident was a deliberate lie, a self-deception or a fantasy, but I did not challenge it.

Shortly afterward I visited Goroka and read the hospital's file on Hari. The events directly leading to Hari's incarceration, according to the file, began with an appeal by the doctor at Tambul to the Assistant Provincial Community Development Officer to do something about Hari, as the Tambul medical unit was unable to cope with him. After interviewing Hari's wife the community development officer informed the Medical Superintendent of Mt Hagen Hospital in writing that Hari 'goes around chasing people with an axe, cutting down trees, killing village pigs and putting on public shows. This has been going on for the past five years'. The head psychiatric nurse from Mt Hagen Hospital subsequently visited the valley and spoke to Hari, concluding after a thirty-minute interview that Hari was 'seriously mentally ill'. He collected tales of Hari's violent behaviour from people in the community, and noted that Hari denied there was anything wrong with him and claimed the community's stories about himself were false. No action was taken at the time, but some weeks later, after a brother of Hari's wife claimed Hari was getting worse, Hari was taken to Mt Hagen Hospital by police. From there he was transferred to Goroka Hospital, five hours' drive from Mt Hagen and two provinces away from his home.

He was not violent or aggressive at Goroka and was put in a normal patient's room (which had a locking door and a window with a light security grille) and classified as 'paranoid schizophrenic'. It was intended that Hari would be put on a régime of chlorpromazine. However, the day after he arrived he tore the grille from the window of his room and escaped. No attempt was made to apprehend him. The file's portrayal of Hari as a chronically violent madman did not fit my experience of him. I was shown the room in which he had stayed. The window still bore the marks of his exit. I imagined him hooking his fingers in the wire of the aluminium grille and wrenching it away from its fittings. No superhuman strength would have been needed.

Returning to the valley I intensified my questions among the community about Hari's alleged violence. Descriptions of his outbursts followed a stereotype which I found was always present in local stories of extreme and florid male madness. He 'destroyed gardens', he 'killed a pig and ate it raw', he 'chopped down trees for no purpose'. The Kakoli were almost exclusively dependent on garden produce for subsistence, and valued pigs as manifestations of wealth, prestige and a rare protein resource whose cooking feeds ancestors as well as the living. They were also aware that the forest that they exploited for game, extra food, building material and firewood was inexorably diminishing. The stereotypic behaviour attributed to Hari amounted, then, to a transgression of basic Kakoli mores. Yet I could find no-one who had actually witnessed any of it. Then I was told of another example of Hari's violence: He had burned down the house of a European linguist in the area. The linguist, from the Summer Institute of Linguistics (SIL), had worked in the area for a number of years. He had operated for a period from a house in the vicinity of Hari's home, and had since left the area indefinitely, according to local people. In his absence the house had been burned down, allegedly by Hari, who had stolen its contents, including clothing which he had since been seen wearing.

The linguist and his wife were eventually to break an impasse for me in my efforts to demystify Hari's persona and legend. Meanwhile I was making little progress. The tax census roll of 1968 at Tambul gave Hari's birth date as 1951 (an approximation), which meant he would have been aged about twenty-two when he went to Bougainville (according to the date he gave me) and about twenty-eight when he returned. At the time of my research he was in his mid thirties. A prominent older man from his home place told me again that Hari had gone crazy after returning from Bougainville and had been violent. The older man, influential in the community, had wanted to have Hari sent to Port Moresby for treatment (i.e. to Laloki Hospital, the country's only exclusively psychiatric hospital) but was a member of a clan who were traditional antagonists and intermarriage partners of Hari's clan. If anything had gone wrong with the venture (i.e. if Hari had met with misfortune or death) it could have sparked interclan friction or even warfare, so for diplomatic reasons the idea lapsed. When I pursued the question of Hari's violence with the community it was conceded that, apart from the burning of the linguist's house, Hari had not been seen to be violent for some time, but I was assured that he had been violent when he first returned from Bougainville.

By now the social relationship between Hari and myself was cooling; he was less enthusiastic toward me than in our first few encounters. I thought this might partly be because I did not follow a particularly 'European' lifestyle, according to popular conceptions of European behaviour and to Hari's experience of Europeans in urban settings. He had expressed surprise that I was living in a 'traditional' house, and I had discovered indirectly that many Kakoli wondered why I did not dress in what they considered the usual European male way (neat casual clothes, rather than my torn t-shirt and shorts which were similar to the attire of many Kakoli men). Visitors to my house also remarked to me that I did not have much *kago* (Tokpisin: material goods). In these respects I differed from the linguist, from the Catholic priest at Kiripia mission, from occasional male European visitors to the valley, and certainly from most Europeans the Kakoli would have encountered in towns. If exclusive familiarity with Europeans was important to Hari I was probably a disappointment to him. But I also suspected that Hari was becoming apprehensive of my slightly probing questions.

I had never seen signs of violence in Hari, but an incident occurred one day that hinted at the kind of behaviour which could generate the image of a dangerous giant. I was writing in my house late one afternoon and had closed and latched my door, which was the only way to ensure privacy if I wanted to concentrate on writing during the day. I heard my name spoken and an instant later the door flew open, the latch disintegrating in the process, and Hari was suddenly inside my house. He was extremely agitated and barely coherent. He had been to see the agriculture officer at Tambul about starting a commercial potato-growing project, but did not understand what had transpired. There had been talk about money and he had been told to see the ADC. As he spoke he was pacing clumsily round the tiny room slapping at the walls of woven *Miscanthus,* causing the house to shudder and clouds of soot[3] and dust to whirl in the air. I could imagine the intimidating effect of this huge, stuttering and thrashing figure on people around him. Hari was simultaneously in panic and despair, stammering in English and Tokpisin that he was just a *longlong* man, he did not understand what was going on. This was the first time I had heard him refer to himself as crazy.

I guessed that the agriculture officer had probably referred Hari to the ADC for purely formal reasons connected with applications to start projects of this nature. Lamely, and unthinkingly, I suggested that Hari simply go to the ADC and get clarification. This incited further battering on my walls. 'No! I can't do that!' he raged. I reminded myself that in view of the problems with his marriage, and possibly his admission to Goroka Hospital, interaction with Tambul authorities like the ADC was probably traumatic for Hari. His confusion over the conversation with the agriculture officer was doubtless exacerbated by his other problems in this respect. I told him I would go to Tambul in the next day or two and find out what was going on and let him know. This calmed him a little, and he soon left. His presence and demeanour had alarmed my neighbours, and I was told emphatically after he left that they did not want him to come visiting me. They were unimpressed by my protestations that Hari was less of a threat to them than they imagined.

The agriculture officer (a European) seemed surprised when I told him of Hari's confusion and panic. Hari, he said, had come to ask about the office's incentive scheme for commercial cash cropping in the area. The scheme involved an issue of seed potatoes, which would be repaid from the first crop in the project. Hari had simply been asked to see the ADC to get his potato-growing area surveyed and officially documented, and to see the Tambul business development officer to arrange for a small loan to help get the venture underway. Clearing the land and planting the potatoes would be Hari's own responsibility, and the office would give him practical advice when needed. During our conversation the officer volunteered to see the ADC on Hari's behalf, and I urged that Hari be encouraged in the scheme as it might be generally therapeutic for him.

I explained the situation to Hari the next time I saw him; he received the information as if he were already aware of it. While we were talking we were passed by men in ceremonial dress on their way to a fundraising *singsing* (Tokpisin: dance, party, musical celebration, etc.) mounted by a local clan to amass money for a compensation payment. I realized that I had never seen Hari in a vestige of 'traditional' dress, not even cordyline leaves. Cordyline was planted extensively around domesticated territory and traditionally had ancestral associations. In past times Kakoli men wore its leaves attached to the back of their waistbands, covering their buttocks. By the 1980s this was still common, and although many adult men had adopted shorts or trousers, they frequently wore cordyline as well, displaying an enduring sense of connection to place and *tamana* (ancestors). I asked Hari if he ever wore traditional or ceremonial clothing. To his negative reply I commented that perhaps it was because he could not be bothered. No, it was not that, he said, it was because he was not treated as part of the community. No one ever talked to him, he was always left alone; this was why he never decorated himself or joined in traditional activity.

While small incidents and conversations like these were giving me insights into the way in which Hari's 'outsider' role and madman persona were developed dialectically by himself and the community, I was still frustrated in my attempts to understand the background to the contemporary situation. It was at this juncture that the linguist and his wife (also a linguist, but apparently not recognized as such by the Kakoli) returned to the valley, not to the area where they had been previously, but staying temporarily at a disused Lutheran mission house at Alkena (across the valley from my host clan's area) and then moving to a disused administration staff house at Tambul. In due course I met the pair and in general conversation I discovered that they had known Hari before his trip to Bougainville: In fact, he had worked for them on linguistic projects in the early 1970s. At that time he had been, according to them, 'a fine figure of a man', standing out from everyone else, highly intelligent and an excellent worker on their translation projects. He had apparently spent a year at school but had been 'too old' to stay longer, though he learned to read reasonably well. The linguists had been sufficiently impressed with him to take him to Ukarumpa, in the Eastern Highlands Province, a PNG headquarters for SIL linguists where he worked in 1972 and 1973 (as a majority of SIL operatives are American, Ukarumpa may have been the source of Hari's gloss 'Americans' for Europeans). Hari had been a good-natured man at that time.

He had then gone to Bougainville and subsequently to Port Moresby. His elder brother heard that Hari had died in Port Moresby and went to investigate. He found Hari alive, but recovering from a near-fatal illness. The linguists did not know the nature of the illness (and I was never able to locate the elder brother, described as a 'businessman', who seemed to be absent from the valley for the duration of my fieldwork period). The linguists had been in the valley when Hari returned from Port Moresby. He told the linguists that he had suffered severe headaches while away, which he described as 'pressure in his head', and had sought medical attention. He alleged that a doctor had prescribed cannabis: the linguists did not know whether to believe this or not. After his return to the highlands in 1979, Hari took a job in Mt Hagen. He suffered a severe headache while working one day and lost consciousness. When he recovered shortly afterwards he learned that he had had a violent fit while unconscious. This occurred several times, and he told the linguists about it in distress and sought medical advice (the linguists did not know what had transpired from this). Between the fits, they said, he had been normal. He had married in 1979 and had been 'devoted' to his wife.

Hari had returned from Bougainville and Port Moresby with ideas for developing communal cash-cropping schemes and improving the community's economic lot. While the linguists had regarded Hari's ideas as sound and probably workable, the community had been apathetic and unco-operative and nothing ever came of his efforts. Hari had also developed strong ideas about 'cleanliness' and 'hard work' during his absence from the valley. The linguists had then left the valley for some time. They said that the flamboyant and nervous behaviour Hari was now exhibiting was a development of recent years. The linguists had heard the tale that Hari was responsible for the burning of the house in their absence, but did not know whether Hari was really the culprit. Hari had been seen at the house while it was in flames, and had apparently attempted to keep people away from it, but whether this was an attempt to stop fire-fighting efforts or to save people from harm was unclear. The linguists were hoping to re-establish contact with Hari and find out his version of the fire incident.

The new information reinforced most of my intuitions and made some sense of Hari's present demeanour and position in the community. It provided likely sources for his chronically violent reputation, and the sophistication of his English (from the SIL community at Ukarumpa and probably the seeking out of European company at Bougainville). I postulated that the debilitating fits and the community's reaction to them must have been a drastic blow to Hari's self-confidence after his experiences as the linguists' protégé in the valley and it was not difficult to imagine how this, coupled with his frustration at the community's lack of response to his ideas about development and the 'work ethic', had contributed to his present condition and relationship with his own society. From the linguists' information it seemed that the violent fits had stopped some time ago, but Hari's reputation for violence was possibly being kept alive by occasional incidents like the hysterical reaction he displayed at my house.

Knowing now that Hari had suffered some kind of illness at Port Moresby, and with sympathy for his situation, I allowed myself to consider his symptoms ethnocentrically. I considered the possibility that the violent fits after his return to the highlands and his ongoing dysphasia could share a single neurological origin. If this were so, it raised the possibility that Hari could be helped pharmaceutically: a mild medication of an anti-epileptic nature, for instance, might lessen his dysphasia and lower his general tension. If nothing were done I expected that Hari would deteriorate gradually, as a result of the social relationship between himself and the community, if not of the long-term worsening of any neurological or psychomotor damage he may have suffered. I needed to know more about the Bougainville and Port Moresby episodes, particularly the near-fatal illness in Port Moresby.

I decided to approach Hari directly, now that I had some reasonably definite information to guide me. This raised problems, and caused me a deal of reflection on questions about the boundary between participant observation and active interference in local sociality. Hari was suppressing information (assuming that he was not sufficiently damaged as to have genuinely forgotten) and would probably not welcome its being recovered. I had to consider the possibility that I might make matters worse: After all, things had been fairly quiet for him since the Goroka Hospital admission. Secondly, there was no guarantee that any information Hari chose to impart would be trustworthy. Hari had misrepresented the trip to Goroka, and on other occasions had volunteered stories about himself which had proved to be untrue. He told me, for instance, that he could play the guitar and sing 'American' songs; 'beautiful music, people cry'. On one occasion in the Kiripia marketplace when he was offered, in my presence, a guitar to play (with malicious humour), it was demonstrated that he was unable to play at all. Whether Hari expected these kinds of claims to be seriously believed was hard to tell; they may have been idle fantasies for his own amusement or benefit. At any rate, they played havoc with attempts to get useful information from him (I considered, too, the possibility that this was deliberate). Since my intentions amounted to a definite intervention in Hari's future social existence (if I should discover that there was indeed some neurological dysfunction which could be overcome), an important third consideration was that Hari might not want to be 'helped'. I had to face the possibility that what I intended as therapeutic intervention might be unwelcome interference.

On my next encounter with Hari I mentioned that I had met the linguists. He received this news without the enthusiasm that he usually displayed when Europeans were mentioned in any context, and he seemed guarded. I said I understood that he knew the linguists well. He affirmed this and changed the subject. With the passage of time, not only had conversation become less easy with Hari, but he had also begun to demonstrate fairly graphically that he was losing interest in my running companionship. We had developed a convention of running back down-valley from Tambul if we happened to meet there, but now Hari would increase his pace up the first hill, leaving me far behind, and disappear at withering speed (the exaggerated tales of his running feats seemed less outrageous now; he really was an outstanding runner). He may have been simply tired of my mediocre pace, but it seemed to reflect our increasing social distance at the same time.

During this period a returning migrant of the valley told me that he had been in Port Moresby at the same time as Hari. Hari had been working as a security guard for a commercial firm, and had not been *longlong* at that time. The migrant could tell me nothing about Hari's sickness however; he had left Port Moresby while Hari was still working. This fragment of information made the illness in Port Moresby focal for me in respect of Hari's 'madness'. I mentioned to the medical officer at Tambul that Hari had apparently almost died in Port Moresby and there was a possibility that it could have some bearing on Hari's behaviour. The doctor said he would try to find out more.

The linguists had shown concern for Hari's condition and expressed interest in contributing to any therapeutic endeavour. However, it soon became clear to me that Hari did not want to re-establish a relationship with them. He made no effort to visit them alone when he was at Tambul and only if there were other people present would he approach their house, drifting through the group and away again. He would converse on these occasions but the linguists were denied the opportunity to speak to him alone. Meetings between myself and Hari usually occurred at Tambul or Kiripia and by this time had ceased to draw an audience. This was partly because I was now a familiar (though still curious) figure and partly because Lopa (see chapter 5) was now providing a theatrical diversion and amusement daily at Tambul. While Hari was displaying some reserve toward me now, he was always affable and I was still the only person with whom he would converse to any extent. He would exchange greetings with kinspeople, but such interactions would not go beyond basic courtesies.

During one encounter at Tambul while Hari and I stood talking and watching a card game (cards were a major preoccupation in the valley, particularly among men) I asked him, without preamble, whether he had ever suffered from bad headaches. My question precipitated all the symptoms usually brought about in Hari by the attention of a crowd, such as the nervous hand movements, the dance, the increased dysphasia. I half expected him to leave, but he spoke briefly about the subject, probably realizing that I had learnt something from the linguists. Yes, he said, he had suffered bad headaches in Bougainville and had left Bougainville as a result. He had seen doctors, who told him there was nothing physically wrong with him. He had also experienced headaches when he returned to the highlands. These had been different headaches, much worse – they made him lose consciousness; he had seen doctors at Mt Hagen and Mendi about them. But that was all in the past, the headaches no longer occurred. He was emphatically dismissive: He felt fine now, there was nothing wrong with him at all. I persisted a little; did he know what had caused the headaches in Bougainville? No, he said. Then he remarked that he had 'had some *wari*' in Bougainville, but did not want to elaborate. He would say no more, and I questioned him no further as he was very uneasy. The reference to '*wari*' in Bougainville was frustratingly vague. *Wari* is a Tokpisin transliteration of the English 'worry', but has wider application; he could have meant anything from anxiety to violent confrontation of some kind. The latter would have raised the possibility, for instance, of the headaches being connected to head injury. He had not mentioned being ill at Port Moresby and I wondered whether by 'Bougainville', he had really meant Port Moresby, but under the circumstances I could not pursue the matter.

The conversation brought into relief for me the ambiguities of my therapeutic impulse. Hari had always insisted that there was nothing wrong with him, except on the occasion when he came to my house after his visit to the agriculture officer. It was obvious that he did not welcome my disinterring episodes of his life that he had tried to bury. He seemed to want to live with his problems as best he could. I did not feel that I could harass him for information he did not want to impart, and I doubted that he would welcome any pharmaceutical or other medical help even if my suspicions about his neurological state were to prove correct. Since my probing was discomforting Hari and would probably cause him distress if I continued, and since his denial of any problem meant he would almost certainly refuse 'help', I decided to abandon any further questioning of him on the subject. Moreover, because of his attitude, I felt obliged to abandon any ideas of 'helping' him, in favour of straightforward observation for the rest of my fieldwork period.

A few weeks after this resolution the doctor at Tambul told me he had managed to ascertain that Hari had suffered a bout of meningitis in Port Moresby and had been critically ill. The doctor had no further details, but the general term 'meningitis' connotes, in adult cases, a wide range of possible side- and after-effects, including epileptic seizures and dysphasia. More detailed information on the specific form of meningitis Hari had suffered would have provided a reasonable guide to the particular damage sustained.[4] As it was, the doctor's information pointed to the possible efficacy of medical intervention. Subject to clinical examination Hari might have benefited, for example, from a low-dose régime of an anticonvulsant medication. Such possibilities would remain conjectural, however, in view of Hari's disposition.

During fieldwork I had constantly tried to change the community's attitude toward Hari, but by the time my fieldwork ended, I had achieved nothing of significance through my efforts. I had attempted to dispel the 'madman' image by insisting (without proper ground, of course) that Hari had suffered an illness which had made him behave strangely, but that there was 'medicine' available which could probably help him. This had no impact on the community, since it appealed to European distinctions between organic and inorganic causes of psychotic behaviour and accompanying quasi-scientific distinctions between 'mental illness' and rectifiable psychomotor dysfunction. None of this, of course, was relevant to the community's everyday concern with Hari's size, strength, unpredictability and, in their view, potential for violence. When I left I felt that only brute social evidence of a change in demeanour in Hari would change their attitude toward him.

The Social Construction of *Kekelepa* Hari

At the simplest level of interpretation, the story of Hari raises fundamental questions about the adequacy of the cross-cultural or transcultural project of psychiatry. The circumstances of Hari's admission to Goroka Hospital's psychiatric ward, and the superficial and inappropriate diagnosis (paranoid schizophrenia) demonstrate the impossibility of the initial ideal implied in Burton-Bradley's stipulation that diagnosis should be qualified by recourse to 'anthropological studies, patrol reports, the kinsmen of the patient, mental health records and where appropriate from the patient himself' (Burton-Bradley 1973a: 5). Hari's experience at Goroka also raises questions about the appropriateness of the more pragmatic model of transcultural psychiatry abroad in the 1980s which accepted that the decision about whether a person was 'disturbed' had already been made by professionals, kinspeople and other community agents inasmuch as they had presented a person for psychiatric assessment, so the assessor's task was 'to determine the kind of psychopathology the patient is suffering from rather than establish its presence or absence' (Draguns 1984: 47–48).

The ethnographic investigation of Hari's estrangement reveals that the Goroka psychiatric unit's understanding of the case was misinformed, to some extent deliberately, by members of the community. The ostensible evidence of psychopathology (running, unpredictable violence, strange speech, etc.) is shown to have been at least in need of some qualification in terms of historical social dynamics and sometimes was a matter of contention. Instead of a referential constellation of beliefs and customs confirming or invalidating a proposition that Hari was mad (the imagined recourse of transcultural psychiatry), a historical process of estrangement has been uncovered whose contributing factors included the enlistment of Hari by the SIL linguists, his working sojourns in Bougainville and Port Moresby, illness, his wife's efforts to leave him, his efforts to keep his children, and so on. The narrative indicates that Hari's madness was socially constructed over a period of time; that what the Goroka psychiatric unit treated as an individual psychosis was, in fact, a social process, and that the psychiatric encounter itself was constitutive rather than therapeutic.

Themes developed earlier in this book are implicit again in the account of *kekelepa* Hari: the practical inadequacy of transcultural psychiatry, the dangers of imposing psychiatric classification and rationality on the basis of what is reported to the psychiatrist. In this respect my own unsuccessful appeals to the community for a more sympathetic view of Hari only served to highlight the point that 'mental illness' is a psychiatric concept and did not properly correlate with Kakoli concepts of even serious chronic madness. Hari was mad, as far as the community was concerned, on the evidence of his social behaviour, both real and imagined. The origin of that behaviour in physical illness, cerebral damage or any other 'non-mental' misfortune was beside the point. I had attempted to manipulate the community's attitude with an argument that required the Kakoli to have internalized broad medical-scientific and psychiatric themes, as has occurred in my own society over the past century and a half. Hari was not 'mentally ill', I tried to tell them; his behaviour was caused by a physical illness. It was the kind of rationalization commonly used by medical professionals and social workers in the West to counter community fear and hostility in the face of incomprehensible behaviour. While there were no objections to my proffered explanation of the origin of Hari's behaviour, it was clearly ineffective in changing the community's attitude toward him. In the view of the Kakoli, whether or not he was 'mentally ill', he was certainly mad.

The dialectic of the social construction of Hari's madness is relatively transparent in the narrative. The seizures and the dysphasia which developed after he returned from Bougainville and Port Moresby, along with the constant running, marked his singularity. His attempt to adopt and apply ideas he had picked up from white acquaintances, and the frustration of these, had compounded his estrangement from the community. His reaction to the community view of him as *kekelepa,* such as his 'nervous' responses under the public gaze and his criticisms of the community, reinforced their perception. His custody struggle with his wife and the admission to Goroka Hospital added an official stamp to the notion that he was crazy. Eventually he was inextricably trapped by the general view of him as a madman – he had become an outsider. He behaved like an outsider and had developed a rationalization of his role, historically fed by the 'work-ethic' and development ideas he had absorbed away from the community: He was different; he had no time for sitting around story-telling and gossiping. Articulated into his self-justification was the repression of past events, the denial that there could be anything wrong with him. In this respect his running may have served a therapeutic purpose, but it also seemed somehow allegorical, distancing him from his own society at the same time as it demonstrated his fitness and good health.

But there is an aspect of the narrative of Hari's madness that needs further exegesis. In general, individual madness which I observed among the Kakoli did not automatically result in chronic marginalization of the madperson, nor in exaggeration or myth, though we have seen that the madness of Kapiye, for example, was mythologized by the community. To understand why Hari in particular became the subject of a legend, I believe we need to contextualize the social construction of his madness in a moral imperative grounded in the harsh conditions of production which govern social relations among Upper Kaugel people. In this respect the ethos of communalism which I referred to earlier is, I believe, a significant determinant of the community's attitude toward Hari.

The obligation to share resources was all-pervasive. It extended to those returning migrants who brought cash or goods into the valley, or to well-established urban migrants, who were expected to host kinsfolk who found their way into town. Men from the valley who had become successful in business or national or local administration positions were expected to be generous with whatever resources they had. Circular migration and bi-locality have long been common-place in post-colonial PNG: Popular demographic conceptions dichotomizing the populace into townspeople and rural 'villagers' fail to take account of the constant movement of people around the country, between homesteads, hamlets, large and small provincial towns and cities. While the majority of Kakoli were more or less permanent valley residents in the mid 1980s, there was nevertheless a small but steady flow of migrants leaving and returning, with various intentions and fortunes. Of those who left in search of formal education, wealth or prestige, some returned successful and shared their fortune visibly through favours such as nepotism in employment opportunities, generous contributions to bridewealth, loans of cash, provision of motor vehicles, beer parties or more formally in the use of their acquired skills for the benefit of the community. Others returned with nothing but tales of adventure or frustration in a distant town and worked to reintegrate themselves into familiar social relations.

The linguists' patronage of Hari in the early 1970s had been portentous. While there had been missionaries and other Europeans in the Kaugel Valley in small numbers since the 1950s, quasi-personal relationships with Europeans were to remain rare for some time. To be taken to Ukarumpa, a European enclave in the Eastern highlands planned along the lines of an idealized small American town, in 1972 was unusual and made Hari singular in his community. The dates Hari gave me covering his absence from the valley fitted with those provided by the linguists. Hari returned in 1979, boasting of his 'American' connections and his marijuana use in Bougainville, seven years after he had left as an SIL protégé. But despite this self-aggrandizement he failed to make any visible social or material contributions or to provide opportunities for the community to benefit from the status he appeared to claim.

Hari's ideas about communal cash cropping were, contrary to the linguists' opinion, unlikely to be viable and the community would have had good grounds for apathy. Clans and clan sections in the upper Kaugel Valley did not all share access to the same types of gardening land (see Bowers 1971: 17) and individual gardens were the subject of complex manoeuvring and disputes over rights and access. The production and exchange of subsistence crops was a matrix of clan, lineage and marriage relations. Individual cash-cropping ventures had mostly been chronically frustrated despite the efforts of successive officers at Tambul's agricultural department station to provide assistance and incentives. The failure of cash cropping was due not only to difficulty in getting goods to markets in Mt Hagen, but also to problems encountered in extracting garden areas from the subsistence network. Multiple claims on specific garden areas and their produce, the planting and nursery strategies of women (who did most of the gardening), problems of vehicle access for bulk carrying and other considerations meant that individual cash-cropping plans were liable to become bogged down in negotiations within the community.[5] A communal cash-cropping project, where more than individual gardening areas were utilized, would have required unprecedented planning and would have been potentially disruptive of social relations. Already frustrated in their attempts to develop stronger links with the cash economy, the Kakoli may have found Hari's proposal an aggravation.

To this was added insult in Hari's portrayal of the Kakoli as lazy. Many 'pre-contact' activities such as large-scale warfare and pre-Christian religious rituals had diminished significantly, as they had in other highland areas. The effects of this diminution were complex. For example I was surprised during my fieldwork to find that younger people were ignorant of the complicated formal counting system in their own language, and mature people had great difficulty in remembering it. Yet it had been recorded by Bowers and Lepi only a decade previously (1975). The distributions of goldlip pearlshells, pigs, game, and other valuables to which the counting system had been linked (ibid., 312–14) had diminished or been transformed by the addition of cash and beer to a degree which had resulted in the atrophy of a system which, as Bowers and Lepi pointed out, had the capacity to be used logarithmically and for dealing with 'international finance or abstract functions' (ibid., 321).[6] In fact in a discussion of counting in 1985 with 'knowledgeable' old men, I found myself (a novice in Umbu Ungu) in the peculiar position of prompting them beyond the number '8' with the aid of a copy of the decade-old paper by Bowers and Lepi, which I had brought to the field to improve my own language facility. The disappearance, one way and another, of a range of institutions together with the hindrance to possible involvement in wage employment of the valley's relative isolation had certainly led to a lack of regular physical working activity (predominantly among men, as women worked steadily in gardening, child care, pig care and other regular activities). The visible inactivity among males was to a large extent unavoidable, and a source of frustration for them, but there was no justification for portraying them as lazy. My own observations were that when there was something to be done (garden clearing, tree felling, house building, etc), people worked hard and willingly, and that young adult males wished there was more work available (gender attitudes appeared to preclude them from helping women in the gardens, though elderly males often kept themselves busy with gardening). Trips to Mt Hagen or further afield in search of work and adventure were not uncommon.

Hari, a physically large and ostensibly robust man, favoured by the linguists, had returned from a seven-year absence boasting of his social accomplishments, but with nothing to offer the community but an unrealistic scheme about cash cropping and unfair censure of a condition they could do nothing about. Compounding this disappointing outcome after a promising departure from the valley, he had 'gone mad' when he returned and possibly, during the early period of distress described to me by the linguists, his violent fits had rendered him an unpredictable liability. Had he represented himself on his return as having failed to make good, as many other hopeful migrants did, he might have received sympathy, but his self-aggrandizement and his condemnation of his fellows negated the possibility of a modest reintegration into the community.

Not only was he rendered a social isolate, but his size, strength, madness, dysphasia and penchant for running lent themselves to caricature and to the construction of a mythological figure. This, I believe, is the key to the exaggerated persona that was presented to me at the beginning of my fieldwork. Hari had transgressed basic codes of behaviour by presenting, and continuing to present, himself as a man of high social stature[7] while failing to fulfil the social obligations that were perceived by the community as a necessary corollary of such stature. The construction of *kekelepa* Hari was a communal exercise in moral iconography: Elements of his appearance and behaviour were exaggerated and fused to create a giant who was mad. Discursively, the Kakoli perpetuated *kekelepa* Hari as a creature physically and socially alien to the community, negative testimony to the importance of aspects of social conduct regarded as morally fundamental. His utility as an icon, imaginatively created from the elaboration of observable characteristics, made the community relatively impervious to my attempts to change their attitude toward him. My naïve project had been an attempt to rehabilitate a man ostracized because he was mistakenly (I thought) believed to be dangerously mad: I had failed to apprehend his significance to the community as an objective reminder of a fundamental moral imperative.

Notes

1. The situation is very different now. Marijuana is grown in many parts of the highlands and widely used.
2. This was not an unusual distance to walk, in fact. Before the increase in vehicular opportunities Kakoli conventionally walked to get to Mt Hagen, stopping overnight on the way.
3. Houses were warmed by a central fire. With no chimney or windows, this meant the upper parts of a house's interior became caked with soot, which improved insulation and reduced infestation by insects.
4. 'Meningitis' is a general term covering many forms, with both short-term and long-term effects, and is usually diagnostically qualified.
5. Nancy Bowers (pers. comm., 1995) noted, though, some short-term cash-cropping ventures instigated by employees at government institutions at Tambul, using land beside the road (i.e. easily accessible for vehicles) in the upper valley. In these cases the Kakoli in effect supplied the land and labour on a temporary basis.
6. Schools taught English counting, and Tokpisin counting was used in the valley, the latter becoming progressively Anglicized with the passage of time.
7. I am deliberately avoiding the overworked term 'bigman'.

CONCLUSION:
IN ANTICIPATION OF A
KAKOLI ETHNOPSYCHIATRY

A critical theme in the early part of this book was the ideological nature of psychiatric theory which has served to legitimate a form of social control by representing the latter as a therapeutic concern for the mental health of individuals. This perspective informed my description of the formal introduction of psychiatry in PNG. Psychiatrists in the 1950s argued that there would be an inevitable rise in madness under the onslaught of civilization, and concern about a generalized 'stress of civilization' legitimated the institutionalization of psychiatry in the late colonial period. The weakness of the argument (for no practical evidence was forthcoming in the following decades) was subsequently veiled by the development of theoretical discourse on the problems of cross-cultural diagnosis. The polymodal development of the 'cross-cultural' discourse (cultural relativism, organic versus inorganic psychosis, psychosis versus psychodrama, and so on) consequently served to project psychiatry into Melanesian 'culture' in the guise of liberal sensitivity to local contextualizing factors.

The discursive field of transcultural psychiatry at the time of my fieldwork was a spectrum determined by two disciplinary concepts: psychiatry's self-understanding as a Western science, and a simple anthropological notion of cultural difference. The spectrum's polarities were ethnocentrism (the uncritical universal application of a strictly medical-scientist psychiatry) and cultural relativism (in effect, capitulation to the 'native point of view'). The ethnocentric extreme was, of course, never argued. The discourse was usually concerned with an area toward the relativist end of the spectrum. Again, the extreme was never argued (otherwise psychiatry would attack its own legitimacy), but discussion centred on two main themes: the contribution that psychiatric expertise could offer to non-Western societies (see, for example, Kiev 1972; Pedersen, Sartorius and Marsella 1984); and the problem of separating mental disorder from cultural norm (see, for example, Burton-Bradley 1973; Kiev 1972).

The theoretical prescription appeared straightforward – if one could understand the culture, then the psychotic (as deviant or defaulter) could be clearly perceived. For the psychiatrist who did not understand the culture, the psychologistic and homogenizing implications of the culture concept ('the Kakoli believe....') provided a last resort: Ask the natives whether the individual is crazy. Much of this book has implied that such a strategy was inadequate and could be misleading. It should not be surprising, therefore, that despite continuing self-critique in the discourse of transcultural psychiatry (see chapter 2, this volume) psychiatric practice in PNG has resorted in the twenty-first century to the guidelines provided by the American Psychiatric Association's 'DSM-IV' model (APA 1994), the safety of 'modern' diagnostic categories, and a training programme for local psychiatrists drawing on conventional Australian practice. In the postcolonial period, while Melanesians have extensively replaced Europeans in professional practice, they have not brought with them a locally informed 'cultural' critique of Western psychiatry. On the contrary, they appear to have found its orthodoxy more comfortable than a search for local 'cultural' diagnostic contexts (see, for example, Koka, Deane and Lambert 2004; Orotoloa, Williams and de Moore 2006).

If we can speak of a liberal orthodoxy in the anthropology of madness, I think it is reasonable to suggest that the idea of ethnopsychiatry is representative of its recent form. My contention that the Kakoli did not, when I conducted fieldwork, have a concept of 'mental illness' positions me against ethnopsychiatry in the interests of pursuing an adequate phenomenology of their lifeworld. I have attempted to rigorously avoid attributing to the Kakoli a preoccupation with the 'mental' in their understanding of madness. Their category of *kekelepa* behaviour – behaviour that was socially 'out of place,' whether innocuously or seriously – did not commit them to the notion of a deranged mind located in a brain. Therefore a comparative anthropological category such as 'ethnopsychiatry' which capitulates to the implicit mentalism of psychiatry is primarily inappropriate.

The terms of reference according to which madness was socially constructed and understood by the Kakoli were so broad that I could not represent this book as a study of madness in any way that implied that individual psychopathology or sociopathology was the analytic focus. The application of a perspective oriented to the proposition of an ethnopsychiatry would have misconstrued the episodes described, for example, in chapters 4, 5 and 6: Madness became instead an ethnographic reference point for a more phenomenological understanding of the Kakoli lifeworld. The praxis manifest in these episodes is focal: A dialectical frame retains our sense of the Kakoli as a people 'with history' processually engaged with their immediate world, an ongoing project which is mutually transformative.

The story of Kapiye's madness (involving possession by his dead mother) was a drama of retribution for moral transgression, contextualized in the ethos of communalism and its obligations, in the face of the individualistic implications of engagement with capitalism. The Kakoli explanation of Wanpis's madness, relating it to menstruation taboos, was a reflection on a social problem in the context of influences like the cash economy. It was a complex elaboration of a cautionary tale combining traditional ontology and contemporary socioeconomic conditions. Something similar was involved in the community's imagery of the mad giant in chapter 6: Hari's transgressions, combined with several idiosyncrasies that could be creatively elaborated to construct a dangerous madman, made him negatively representative to the community of the fundamental moral imperative implicated in the story of Kapiye.

In chapter 5 my encounter with Lopa's allegedly cyclic madness (superficially corresponding at first reports to the psychiatric concept of manic depression) was itself part of a singular episode in the praxis of the community. To frame his behaviour in a synchronic 'cultural' context as a repetition of his past episodes of madness would have been misleading, and not simply because my own presence became part of the dynamic of its course. Ethnographically, as a narrative of a contemporary social problem, it illustrated the ways local government, medical services, and other elements of the post-colonial state had become incorporated into local praxis. Episodes of madness (*kekelepa*), then, have been understood in this book as part of Kakoli praxis, the contradictory engagement of subject and object which is simultaneously a surpassing and an internalization. And thus the Kakoli continue to change – which requires some reflection here.

It is more than two decades since I carried out my fieldwork among the Kakoli. I have conducted research in other parts of PNG periodically since 1990 and keep abreast of developments in Melanesia generally, but I cannot give an ethnographic account of whatever developments have taken place specifically in Kakoli ideas about, and attitudes toward, madness. I could not say whether the notion of *kekelepa* – behaviour that is socially out of place – still encapsulates madness generally in the upper Kaugel Valley. Certainly, given the persistent intrusion of Western institutions and rationales into PNG sociality as a whole, contemporary discourses of madness would be likely to reflect newer Kakoli preoccupations than those embedded in the narratives in chapters 4, 5 and 6. Moreover, changes in discourses of madness would not, of course, have occurred in phenomenological isolation from others to which I have occasionally referred, such as the disappearance by the time of my fieldwork (when younger generations of Kakoli were counting almost entirely in Tokpisin) of various practical activities to which the Kakoli mathematical system was formerly linked. Neither can we ignore (though I cannot possibly assess it) the impact and cumulative effect of inexact translations and the appropriation of foreign language terms in general on the Kakoli. During my fieldwork period literate young people were beginning to read missionary Umbu Ungu literature, whose choice of local terms in translation of European ideas seemed to me sometimes equivocal, but the effect of this on Kakoli conceptualization and thought overall was not clearly evident.

One implication of the previous paragraph is that if there has been a shift towards a psychologistic interpretation of some *kekelepa* behaviour – that is, a shift towards an ethnopsychiatry – it would need to be understood as part of wider transformations in Kakoli ideas of the person and of their cosmology. These transformations would not simply reproduce those which occurred in the West, where the development especially of Christianity and capitalism increasingly individualized the human subject (Dumont 1986) and ideologically framed the gradual integration of the soul and the mind. The complex relationship between the soul and the mind – and between the troubled soul and the deranged mind – changed over a period of centuries in European societies. Evidence of the early identification of madness with the soul in the Judeo-Christian tradition can be found in the accounts of tormented souls in the Old Testament of the Bible.[1] A discernable development in the New Testament is a connection between the entry of evil spirits into ordinary men and the workings of their minds.[2] By the end of the eighteenth century therapeutic strategies no longer involved an engagement with the soul, or with supernatural forces, but a dualism of body and soul remained implicit in discourses of madness which were becoming increasingly medicalized (Doerner 1981: 24–25, 100–03). In the age of industrial capitalism, the dichotomization of the individual human into physical and mental aspects was effected partly through the rise of medical science in the nineteenth century. Whether the soul was subsumed into the mental aspect or displaced by it was, for a while, equivocal – partly reflecting the interplay of religious and secular approaches to treatment during that period (see, for example, ibid., 86; Foucault 1982: 255–59, 2006: 491–503; Tuke 1964). However by the late nineteenth century a definite transition had occurred in the conception of madness – moral insanity had become mental insanity, and new terms for madness such as 'mental illness' and 'mental disability' (or for example *aliénation mentale* and *maladie mentale* in the French case) gained currency (Cooter 1981; Doerner 1981: 88–91, 140–49; Scull 1979: 125–63).

Specific historical processes, then, have brought about individualistic personhood and contingent physical-mental dualism in Western society, and it would be naïve to assume that this centuries-long process would simply be reproduced in an accelerated form in societies such as the Kakoli through a few decades' experience of Christianity and a degree of capitalism. On the other hand the complex effect of colonial experience cannot be denied. The end of direct colonialism in PNG and the advent of political 'independence' in the 1970s was not accompanied by any lessening of Western influences. Indeed the independent state of PNG was constitutionally declared a Christian nation and has since become encompassed and colonized by processes more global that those experienced under Australian rule. The various effects of Christianity in particular among diverse groups in PNG have been discussed in a growing body of anthropological literature in recent decades (see, for example, Eriksen 2005; Otto and Borsboom 1997; Jebens 2005; Robbins 2004; Tuzin 1997; Van Heekeren 2004). Direct Christian influence reached the Kakoli in the 1950s – later than it reached several nearby peoples such as the Melpa of Mt Hagen, where missionaries established themselves in the 1930s. By the 1980s a majority of the Kaugel Valley people among whom I worked were nominal members of one or another denomination. Yet despite their participation in Christian rituals, and an anticipation by some people of a millennial visit to the valley by Jesus Christ, there was not a great deal of evidence at that time that their cosmology had been dramatically altered structurally by the incorporation of the Christian God, as I noted in chapter 3.

My interlocutor Manenge was one of a number of older men in my host clan who had not capitulated to Christianity, though his wife was a nominal convert to Catholicism and his children were Sunday churchgoers. His discussions with myself on subjects such as the nature and workings of *numan* and *mini* and more generally on Kakoli cosmology were devoid of references to God. Manenge's explanations were echoed by some other non-Christian informants of his generation who were patient enough to discuss the same topics with me repeatedly and in depth, though these described themselves as having forgotten some aspects consequent to the disappearance of a number of pre-Christian activities under the force of colonial officialdom. This lack of obvious Christian influence in local exegesis of concepts such as *numan* and *mini* contrasts with an exposition by one of Andrew Strathern's Melpa informants on *noman* and *min* (the Melpa equivalents of *numan* and *mini*) six years earlier in 1979 which was clearly already imbued with Christian ideas (Stewart and Strathern 2001: 114–25). As Stewart and Strathern indicate, Christian theological understandings of the relationship between 'breath' and the soul appeared to have had an effect on the informant's formulations, particularly of *min*, which he discussed as if it were equivalent to 'breath' and occupying the role of soul in a Christian-like dichotomy of body and soul (ibid., 116–17).

I heard nothing of this nature in discussions of *numan* and *mini* during fieldwork in the 1980s, despite the nominal Christianity of most Kakoli, and the fact that Christian literature recently translated into Umbu Ungu used '*mini*' as a translation of 'soul'.[3] While God had clearly been added to Kakoli ontology, he seemed to have been attributed a different kind of agency to that of their ancestors and *kuru*, and other entities in their lifeworld. I could only speculate at the time on the hermeneutics of their encounter with Christianity. Among Christian pastors, who were mostly men I judged to be aged in their 30s or early 40s, the Christian message was expounded in terms of the human capacity for 'sin' (locally understood as doing a 'bad thing' – *ulu keri*[4]), the necessity to pray, and the capacity of God/Jesus to help and 'save' the faithful. Despite their pious commitment to Christianity, these men described 'traditional' ontological concepts without resorting to Christian terminology in answer to my questions, as if they were shifting from one distinct discourse to another. They were less skilled in exegesis than older men such as Manenge, and directed me to consult people of his generation about aspects they were not sure of. One pastor, with no apparent sense of inappropriateness to his calling, gave me a detailed account of how a ritual specialist could capture the *mini* of an enemy war leader (thereby rendering him weak in battle) in advance of a skirmish.[5]

Recent literature from the Mbowamb area in general indicates that since my fieldwork period there have been some significant developments in ideas of cosmology and personhood. Among the Melpa, already significantly influenced by Christianity in 1979 as Strathern's informant implied, a range of material processes have been catalogued as influencing changing concepts of gender and personhood, including the growing importance in the 1990s of women in Christian Church activities (Strathern and Stewart 2000: 13). Further, and pursuant to the types of changes in ideas of *noman* and *min* evident in 1979, there have been significant developments in the ontological relationship between the introduced Christian God and Melpa ancestors, whereby God is now seen to be in charge of the ancestors. Nowadays for example, the ancestors have to seek God's permission before making someone sick (2000: 80–81). This, Strathern and Stewart point out, places God the mediator further back in time than the ancestors and also hierarchizes politics. As well as reinforcing a pacifist ideation (the peace of God – the first ancestor – seen to be broken by later ancestors, and requiring restoration), 'the trope ancestor/God turns collective difference into collective unity. God as an individual figure thus collapses all collectivity into himself; he is the ultimate collective individual, we may say' (2000: 81). Strathern and Stewart also note shifts in conceptions of anger, which traditionally was experienced in the *noman*[6] but has come to be associated with the soul under Christian influence. Where the revelation of anger and wrongdoing previously had to do with ongoing social relationships, it now had to do with 'the relationship of the person to God, the future residence of the soul, and its individual mortality' (ibid., 155). We note here again the shift in the local (as compared to the colonial missionary) conceptualization of the *min* – the 'soul', according to the Christian understanding that it is 'the part of the person that survives death and can go to heaven or hell' (ibid.).

More specifically referential to changes among the Kakoli is an educationist's account of 'mother-tongue' schooling in the Kaugel Valley (Malone 2004). The perspective of a missionary linguist is qualitatively different from that of an anthropologist, and Malone writes from his experiences at a school near the valley's head using vernacular teaching methods, drawing on interactions with the school's students and their parents. In translation from Umbu Ungu in Malone's work, *mini* is rendered as 'soul' and his informants apply it in dichotomy with the 'body' (for example ibid., 134). An inference invited by his descriptions is that, at least among the Kakoli who are involved with his school, conceptions of the past are markedly different than they were in the 1980s, and elements of precolonial mythology that were still visible at that time have disappeared. In particular the stories used in the vernacular teaching program described by Malone

– mostly contributed by Kakoli people (ibid., 233–37) including the school children themselves – contain no traces of the cosmological themes which were discernable during my fieldwork period. Interestingly, Malone says that the stories were 'taken from real-life experiences which are well-known in the area' (ibid., 191), and authors 'had the freedom to feature those aspects of community life that they considered of interest and importance to themselves and the children' (ibid., 189). On the ground that the stories commonly have references to, for example, local foods and everyday activities the author argues that they relate closely to local 'culture' (ibid., 177–79).

An example used by Malone is the *Wenewene Temane,* 'An Ogre/Echo Story', which he says provided children with an opportunity to express their knowledge of 'a traditional animistic cosmology that posits a "middle world" of unseen, mostly malevolent beings' (ibid., 178). The story concerns a creature called *wenewene* (ibid., 226–27) and was contributed by a former pupil of the Tambul Bible School. In the story, which is illustrated in the book with comic-strip pictures, some boys make various noises which are echoed by the *wenewene*, pictured as a human-like creature with devil's horns. The *wenewene* (referred to in English both as an 'ogre' and a 'demon') is said to live by the banks of the Ambola River (ibid., 190), which runs into the Kaugel River from the north at the head of the valley.

Despite the author's reference to 'traditional cosmology', however, the echo story as rendered in Malone's book is far removed from the context which had given it meaning in the 1980s and earlier, when a narrative about *wenewene* (which can be translated as 'echo') was part of a complex of origin and cult myths involving Kakoli ancestral figures who migrated into the valley. This complex was itself part of a greater body of myths of the region which are an important guide to pre-European-contact cosmology. Attempts were made by missionary ethnologists of the early contact period to understand this cosmology in a more or less systematic way, drawing on material from the Melpa area and on oral accounts derived from other parts (some unvisited by Europeans at the time) of the greater area of the Mbowamb (Vicedom and Tischner 1943–48; Strauss 1990). Later, in the 1970s, Nancy Bowers and her principal local research assistant, Pundia Lepi, recorded a version of the *wenewene* story as a self-contained narrative and published it in a short-story collection (Lepi and Bowers 1983: 42–45), attributing it to the narrator as a matter of courtesy. It was still being told, as an entertaining and self-contained story, during my fieldwork period but was discernibly related to the mythopoeic lifeworld of successive generations of valley dwellers.

In the versions given to Lepi and Bowers and later to myself the *wenewene* was represented in its primordial condition, as an echo emanating from the volcanic rock at the top of Mt Giluwe. In local cosmology Giluwe's summit had been the place from where an original being had distributed the flora and fauna which later sustained humans in the region. In the story the voice that constituted the *wenewene* was generated by two virgin women who had been transformed into stone after travelling up to the summit during a journey to find husbands. The *wenewene* story made reference also to a dog and a half pig (i.e. it had only one 'side', from nose to tail) who dwelt at Mt Giluwe's summit. These were magical creatures connected to the original being who distributed the life-sustaining resources in the area, and they were said, at the time of my fieldwork, to still exist in Giluwe's crags. Humans who went to the mountain top might accidentally but safely 'see' the half pig from one side, the 'cut away' side, for this was covered with leaves and other bush materials (i.e. from this side the pig blended with the environment), but if they saw the other side, perceiving it in conventional pig form, they would perish. The *wenewene* story, as told in the 1980s involving two virgin women, was integral to the local mythology of the Kakoli which included myths associated with 'goddess cults' in the valley (see Didi 1979, 1982a, 1982b). The latter can be linked thematically to the Mbowamb 'women spirits' in Strauss's 1962 discussion of aspects of the Mbowamb lifeworld (Strauss 1990: 317–23). Further, the transformation of virgin girls into stone was a component of the origin mythology of Mbowamb groups in general.[7] Despite the brevity of my references to 'traditional' cosmology here, I think they are sufficient to show a striking contrast to the *wenewene* story used at the vernacular school in recent times. The latter is devoid of the content I have described despite Malone's own reference to 'traditional' cosmology, and the fact that the story was contributed by a Kakoli man.

Elsewhere in Malone's book a local man misrepresents the phenomenology of past pig sacrifices to ancestral spirits with an empiricist argument about credulity: 'We would sacrifice the pig to the ancestral spirit, but no one came to take the pork from the cooking pit and eat it ... we ourselves ate [the pork] and said that the spirits had eaten it ... All these evil customs are now on the wane' (Malone 2004: 134). We cannot know how much the significant shift in cosmological perspective evidenced in the echo/ogre story and the critical rejection of the possibility of ancestral consumption of pork is representative of the contemporary attitudes of non-migrant Kakoli in general, given the degree to which Christian doctrine is reinforced through the school which serves as the touchstone for Malone's description. Yet it has to be said that in the 1980s among my Kakoli acquaintances even Christian pastors comprehended the difference between the ways ancestral spirits and corporeal humans consumed pork.

An inference to be taken from the observations of Strathern and Stewart above is that there is a difference in the usage of words such as *noman* and *min* (or Kakoli *numan* and *mini*) today, compared to usage in earlier times. But further, inasmuch as *numan* and *mini* are integral not only to conceptions of the person but to the wider ontology of the Kakoli and related groups, it is likely that the ontology which informed the attitudes and behaviour of the Kakoli in the 1980s has since changed significantly. An increasing erasure of cosmomorphic experience itself is implicit in developments like the above-cited transformation of the *wenewene* story from mythic narrative into fairytale and the empiricist incredulity toward ancestral consumption of pork cooked by humans. A Melpa-like shift of emotion among the Kakoli from corporeal locations to a Christian soul (now incorporating the *mini*) would be facilitated by these ontological changes. Notions of affliction, for example, in this changed lifeworld are unlikely to credit a *kuruwalu* (if such entities still exist for the Kakoli) or a dead mother with the ability to take control of a Christian *mini*/soul in the way that the un-Christian *mini* of Kakoli were said to be manipulated in chapters 3 and 4. More generally the gradual decline of the cosmomorphism which gives sense to the complex of relationships among corporeally ordered emotions, sociocentric humans, non-human entities and the Kaugel Valley environment increases the likelihood that post-Cartesian understandings of the person, encouraged by Christianity, capitalism and associated institutions including medical science, are spreading among the Kakoli.

The reinforcement of the post-Cartesian view of the person would undermine any surviving local subscription to a relationship of contingency between thinking (*konopu*) and a socially developed, and socially potent, *numan* (an intersubjective force, rather than a 'thinking' entity) located in a person's chest. It thereby increases the likelihood of the mind (in-the-brain) becoming a psychological locus of individualism, facilitating the perception of madness as an internalized, mentalized condition. Whether technical psychiatric definitions are also accepted into the Kakoli lifeworld or not, the contemplation of madness as a *mental* problem and the possibility of restoring the individual to mental health would constitute a Kakoli ethnopsychiatry. Whether the Umbu Ungu term *kekelepa* would, in the long term, continue to be applied to this mental derangement, is an interesting question.

Notes

1. See, for example, in the Christian Bible, the madness of Saul in the First Book of Samuel 16.14–23, or the torment of Nebuchadnezzar in Daniel 4.31–37.
2. For example Mark 1.23–27, Mark 5.1–15, Matthew 12.43–45, Luke 4.40–41.
3. The Summer Institute of Linguistics produced an Umbu Ungu translation of Luke's Gospel in 1980. However, I did not see this in any Kakoli hands during fieldwork. Some short pamphlets – guides to Christian living, with Bible quotations – were in circulation among young people.
4. This is an example of the difficulty missionaries have experienced in PNG in trying to translate Christian concepts using local language terms. Where Western Christians understand 'sin' to refer to an internal condition in the individual, the term *ulu keri* traditionally referred to bad acts, and was also used to refer to sexual intercourse, which was 'bad' partly because of the dangers associated with leaking bodily fluids (see chapter 4).
5. This man, a regular advocate of 'peace' in his sermons, was killed participating in warfare which broke out after I had finished fieldwork.

6. The authors describe *noman* at one point as the 'mortal mind' (Strathern and Stewart 2000: 155), in distinguishing it from the 'soul' to which the term *min* is applied.

7. In the section just cited (1990: 317–23), Strauss describes what he calls the 'Ngênap Cult', said to have originated in the *Kawudl* (i.e. Kaugel) region. He also discusses the significance of smooth, unchipped stones as representative of ancestral, 'closed' or virginal women.

Appendix A:
Orthography

Umbu Ungu Phonemes

Consonants

	Bilabial	Dental	Alveolar	Velar
Voiceless Stops	p	t	s	k
Prenasalised Stops	mb	nd	nj	ng
Nasals	m	n		
Laterals		l	l	l
Non-nasal Continuants	w	r	y	

Vowels

	Front	Middle	Back
High	i		u
Mid	e		o
Low		a	

Comments

1. The first linguists to work in the upper Kaugel were Bruce and Ruth Blowers, who began gathering material in 1955 and continued until 1964. They used single symbols to represent four prenasalized stops and affricatives: thus [mb] was represented by /b/, [nd] by /d/, [ndz] by /j/, and [ng] by /g/ (for example, Blowers and Blowers 1969; Blowers 1970). The anthropologist Nancy Bowers, who began fieldwork in 1961, represented these complex phonemes diagraphically, thus: /mb/, /nd/, /nj/, /ng/ (for example, Bowers 1968; Bowers and Lepi 1975). My orthography follows Bowers, rather than the Blowerses, partly for simplification (assuming that most readers of this book are not linguists), but also because Bowers' version is closer to what a more recent linguist, June Head, has reported to be the preference of literate Kakoli (Head 1990: 116).

2. When followed by a front vowel, the /l/ is pronounced as a laterally released dental plosive. To a native English speaker untrained in linguistics and unfamiliar

with Umbu Ungu this would sound like /dl/: the word *kekelepa,* for example, would sound like 'kekedəlpa' at first encounter. When followed by a back vowel, it is pronounced as a laterally released velar plosive, which would sound something like 'gl' to the unfamiliar English speaker.

3. While Head (1990) shows /t/ as representative both of the dental stop [t] and of the alveolar [s], apparently following the preference of literate Kakoli, I have distinguished between them. Here I follow Nancy Bowers (1968), and agree with Blowers's inclination (1970: 3) to hear /s/ as an affricated alveolar stop.

4. There are three sub-dialects of Umbu Ungu: one spoken at the valley's head, another in the mid-area (the location of my own partial language acquisition), and another in the lower valley. The differences between them are slight, but may account in part for some of the orthographic variations found reading among Bowers (1968), Blowers (1970), Head (1990, 1993), Bible literature translated into Umbu Ungu (e.g. Anon. 1980), and this publication.

Appendix B:
Glossary of Umbu Ungu Terms

This glossary refers to the usage of Umbu Ungu terms in the mid 1980s. All languages in regular use are dynamic and changing: Some of the terms may now have different meanings, particularly under the influences discussed in this book's conclusion.

alaye	diviner and healer of illness
amu	wild pandanus (*p. jiulianettii*)
amu ungu	'pandanus talk': a coded form of speech traditionally used in the forest during the harvesting of wild pandanus
ambu	woman
ambu kinan	courting ritual in which young couples sit side-by-side, cheek-to-cheek and sing, similar to rituals in other highland PNG groups which are commonly called 'tanim het' in Tokpisin
kakoli	the name used by the people of the upper Kaugel Valley for themselves and for the Kaugel River
kaila	red cordyline
kangi	flesh, soft tissue of the human body
kange	story – usually of the past – told for entertainment (cf. *temane*)
kango	boy
karaye	bend oneself, flex
kel	less, diminished, small (but with potential for growth)
kelepa	to leave a group or a familiar location in order to do something singular or removed from normal practice
kekelepa	mad, crazy (but does not denote mental illness); also used of drunkenness
keri	bad, an act with potentially damaging social or personal consequences

kondo	sorrow
kondole	red
konopu	thinking, thought
konopu talo	'two thoughts', confusion
kopong	grease, semen
kurpule	the appropriation of a person's *mini* (q.v.) by an invisible being such as a dead parent or a feral 'spirit' (*kuruwalu*, q.v.)
kuru	a general term used of invisible 'spirits' of many kinds, and also of illness
kuru tokomo	to be ill: a rough translation is 'spirit strikes'
kuruwalu	a malevolent invisible entity that inhabits areas peripheral to human habitation, a feral 'spirit'
iri	public fury, rage, the visible expression of *mumundili* (q.v.)
makale	a large-scale ceremonial distribution of pigs and other valuables
meme	blood
mini	an intangible component of the person which inheres in the soft tissue of the body, visible to others in a reflection or shadow, or a dream image; its final departure at (physical) death results in the decay of the soft tissue, leaving only the bones ('*mini*' was the term chosen by missionaries as a translation for the Christian concept 'soul')
mini wale	startled, stupefied; refers to the *mini* (q.v.) having momentarily left the person
mongo	eye, centre, trouble (in the sense of a concentration of issues in a central place)
mulu	sky
mulu-kola	sky-place; translation used by missionaries for 'Heaven'
mumindili	a rising anger experienced in the belly
mund	heart
mundemong	'heart trouble': immediate fear (of tangible things), experienced as a tightening in the chest
naa	not, negative case, e.g. '...*naa pilkiru*': '...don't/doesn't understand'
numan	an invisible component of the person, located in the chest cavity; it is a locus of thought, intention and will, able to affect and be affected by others and by relations with others (it has no direct

	equivalent in Western concepts of the person, but has been used in some literature as a translation of 'mind')
olie	shorthand for *olo tekemo* (q.v.)
olo	belly
olo tekemo	'belly-doing': an emotion of inferiority (cf. *pipili*)
ombel	bone
pilkiru	understand, comprehend, know
pimu	blocked up (…passage, tunnel, etc.); also 'deaf-mute'
pipili	a feeling of inadequacy, inferiority, angst, foreboding ('shame' is sometimes used as an approximate but imprecise English gloss)
polu	putrefaction
popolu	frustration anger, gut-rotting resentment felt but not made public (cf. *mumundili, iri*)
Pulu-Yemo	'original root man': term used by missionaries as a translation for 'God'
tambili	name of an area of flat land at the head of the Kaugel Valley (rendered as 'Tambul' in colonial usage from the 1950s and now spelt thus)
tarama	ancestors (collective)
temane	a story of the past intended to explain directly or indirectly some aspect of the present
ulu keri	'bad thing': term adopted by missionaries as a translation of 'sin'
umbu	a term denoting autochthony, also seedling
ungu	language, talk (n.)
ungu iku	'folded talk': utterance with hidden meaning
wapera	promiscuous woman, prostitute
walo	small
wenewene	echo
ye	man
ye kondole	'red man' (i.e. white man)

REFERENCES

Abu-Lughod, L. 1991. 'Writing Against Culture', in R. Fox (ed.), *Recapturing Anthropology.* Santa Fe: School of American Research Press, pp. 137–62.

Allen, B., H. Brookfield and Y. Byron. 1989. 'Frost and Drought Through Time and Space, Part II: The Written, Oral, and Proxy Records and Their Meaning', *Mountain Research and Development* 9(3): 279–305.

Andrew, Br (SSF). 1983. *The Management of Psychiatric Disorders of Patients in Provincial Hospitals in Papua New Guinea.* Goroka: Talair Pty Ltd.

———. 1985. 'Helpim ol Manmeri i gat Sik long Het', in *Wok Sambai* (Christian Institute of Counselling handbook). Goroka: Melanesian Institute.

Anon. 1980. *LLuku: Temane-Peangamo LLukuni Torumu-Bokumu.* Ukurumpa: Summer Institute of Linguistics.

APA [American Psychiatric Association]. 1994. *Diagnostic and Statistical Manual of Mental Disorders,* 4th ed. Washington, DC: American Psychiatric Association.

Appadurai, A. 1996. *Modernity at Large: Cultural Dimensions of Globalization.* Minneapolis: University of Minnesota Press.

Bains, J. 2005. 'Race, Culture and Psychiatry: A History of Transcultural Psychiatry'. *History of Psychiatry* 16(2): 139–54.

Banton, R. et al. 1985. *The Politics of Mental Health.* Oxford: Macmillan.

Basaglia, F. 1980. 'Problems of Law and Psychiatry: The Italian Experience'. *International Journal of Law and Psychiatry* 3(1): 17–37.

Bates, E. M. 1977. *Models of Madness.* St Lucia: University of Queensland Press.

Bateson, G. 1972. *Naven.* Stanford, CA: Stanford University Press.

Billig, O. 1975. 'Foreword', to B. G. Burton-Bradley, *Stone Age Crisis.* Nashville: Vanderbilt University Press, pp. xii–xiv.

Bleuler, E. 1961. *Dementia Praecox, or the Group of Schizophrenias.* New York: International Universities Press.

Blowers, B. L. 1970. 'Kaugel Phonemic Statement', in B. L. Blowers, M. Griffin and K. A. McElhanon, *Papers in New Guinea Linguistics* No. 13, Pacific Linguistics Series A, No. 26. Canberra: Australian National University, pp. 2–12.

Blowers, B. L. and R. Blowers. 1969. 'Kaugel Verb Morphology', in C. L. Voorhoeve et al., *Papers in New Guinea Linguistics* No. 12, Pacific Linguistics Series A, No. 25. Canberra: Australian National University, pp. 37–60.

Bourguignon, E. 1984. 'Belief and Behaviour in Haitian Folk Healing', in P. B. Pedersen, N. Sartorius and A. J. Marsella (eds), *Mental Health Services: The Cross-Cultural Context.* Beverly Hills: Sage, pp. 243–66.

Bowers, N. 1968. 'The Ascending Grasslands', PhD Thesis, Columbia University.

————. 1971. 'Demographic Problems in Montane New Guinea', in S. Polgar (ed.), *Culture and Population: A collection of current studies*. Chapel Hill, NC: Carolina Population Center Monograph No. 9, pp. 11–31.

Bowers, N. and P. Lepi. 1975. 'Kaugel Valley Systems of Reckoning', *The Journal of the Polynesian Society* 84(3): 309–24.

Brandewie, E. 1981. *Contrast and Context in New Guinea Culture: The Case of the Mbowamb of the Central Highlands*. St Augustin: Anthropos Institute.

Bromberg, W. 1975. *From Shaman to Psychotherapist: A History of the Treatment of Mental Illness*. Chicago: Henry Regnery Company.

Brown, P. 1985. *The Transfer of Care: Psychiatric Deinstitutionalization and its Aftermath*. London: Routledge and Kegan Paul.

Burton-Bradley, B. G. 1965a. 'Culture and Mental Disorder', South Pacific Commission Technical Paper No. 146, Noumea, pp. 27–30.

————. 1965b. 'The Psychiatric Examination of the Papua and New Guinea Indigene', South Pacific Commission Technical Paper No. 146, Noumea, pp. 31–32.

————. 1969. 'Papua and New Guinea Transcultural Psychiatry: The First One Thousand Referrals', *Australia and New Zealand Journal of Psychiatry*, 3: 130–36.

————. 1970. 'The New Guinea Prophet: Is the Cultist Always Normal?', *The Medical Journal of Australia* 1(3): 124–29.

————. 1973a. *Longlong: Transcultural Psychiatry in Papua New Guinea*. Port Moresby: Public Health Dept.

————. 1973b. 'The Psychiatry of Cargo Cult', *The Medical Journal of Australia* 2(8): 388–92.

————. 1975. *Stone Age Crisis*. Nashville, TN: Vanderbilt University Press.

————. 1976. 'Papua New Guinea Psychiatry: An Historical Sketch', *Papua New Guinea Medical Journal* 19(1): 1–5.

————. 1979. 'Book Review: Mental Health Services in PNG', *Papua New Guinea Medical Journal* 22: 203–05.

————. 1985. 'Transcultural Psychiatry in Papua New Guinea', *Transcultural Psychiatry Research Review* 22: 5–35.

Burton-Bradley, B. G. and C. Julius. 1965. 'Folk Psychiatry of Certain Villages in the Central District of Papua', South Pacific Commission Technical Paper No 146, Noumea, pp. 9–26.

Bynum, W. F. 1981. 'Rationales for Therapy in British Psychiatry 1780–1835', in A. Scull (ed.), *Madhouses, Mad-Doctors and Madmen: The Social History of Psychiatry in the Victorian Era*, London: Athlone Press.

Bynum, W. F., R. Porter and M. Shepherd (eds). 1985. *The Anatomy of Madness*, Vol 1. London: Tavistock.

Castel, R., F. Castel and A. Lovell. 1982. *The Psychiatric Society*. New York: Columbia University Press.

Champion, I. 1940. 'The Bamu-Purari Patrol, 1936', *The Geographical Journal* 96(3): 190–206.

Chinnery, E. W. P. 1934. 'Mountain Tribes of the Mandated Territory of New Guinea from Mt. Chapman to Mt. Hagen', *Man* 34: 113–21.

CIC [Christian Institute of Counselling]. 1985. Annual Report. Goroka: Melanesian Institute.

Clare, A. 1976. *Psychiatry in Dissent*, London: Tavistock.

Colletti, L. 1973. *Marxism and Hegel.* London: New Left Books.

Cooter, R. 1981. 'Phrenology and British Alienists, ca. 1825–1845', in A. Scull (ed.), *Madhouses, Mad-Doctors and Madmen: The Social History of Psychiatry in the Victorian Era.* London: Athlone Press, pp. 58–104.

Dax, E. C. 1961. *Asylum to Community: The Development of the Mental Hygiene Service in Victoria, Australia.* Melbourne: F. W. Cheshire.

Deleuze, G. and F. Guattari. 1977. *Anti-Oedipus: Capitalism and Schizophrenia.* New York: Viking Press.

Devereux, G. 1956. 'Normal and Abnormal: The Key Problem of Psychiatric Anthropology', in J. B. Casagrande and T. Gladwin (eds), *Some Uses of Anthropology: Theoretical and Applied.* Washington, DC: Anthropological Society of Washington, pp. 23–48.

———. 1961. 'Two Types of Model Personality Models', in B. Kaplan (ed.), *Studying Personality Cross-Culturally.* New York: Harper and Row, pp. 227–41.

———. 1980. *Basic Problems of Ethnopsychiatry.* Chicago: University of Chicago Press.

Didi, B. K. 1979. 'Kuru Wapu Cult in the Lower Kaugel Valley of the Tambul Subdistrict, Western Highlands Province', *Oral History* 7(6): 1–40.

———. 1982a. 'Kuru Kopiaka, Goddess Cult in the Lower Kaugel Valley of the Tambul Sub-District, Western Highlands Province', *Oral History* 10(1): 5–43.

———. 1982b. 'An Overview of the Traditional Religious Cults in the Lower Kaugel Valley of the Tambul Sub-District, Western Highlands Province', *Oral History* 10(1): 44–87.

Doerner, K. 1981. *Madmen and the Bourgeoisie.* Oxford: Basil Blackwell.

Draguns, J. G. 1984. 'Assessing Mental Health and Disorder Across Cultures', in P. B. Pedersen, N. Sartorius and A. J. Marsella (eds), *Mental Health Services: The Cross-Cultural Context.* Beverly Hills: Sage.

Dumont, L. 1986. *Essays on Individualism: Modern Ideology in Anthropological Perspective.* Chicago: University of Chicago Press.

Eriksen, A. 2005. 'The Gender of the Church: Conflicts and Social Wholes on Ambrym', *Oceania* 74(1 and 2): 284–300.

Feil, D. K. 1984. *Ways of Exchange: the Enga Tee of Papua New Guinea.* St Lucia: University of Queensland Press.

———. 1987. *The Evolution of Highland Papua New Guinea Societies.* Cambridge: Cambridge University Press.

Fitzpatrick, P. 1980. *Law and State in Papua New Guinea.* London: Academic Press.

Foley, W. A. 2000. 'The Languages of New Guinea', *Annual Review Of Anthropology* 29: 357–404.

Fortune, R. F. 1963. *Sorcerers of Dobu,* London: Routledge and Kegan Paul.

Foucault, M. 1982. *Madness and Civilization.* London: Tavistock.

———. 2006. *History of Madness.* Oxford: Routledge.

Fox, R. G. and B. J. King. 2002. 'Introduction: Beyond Culture Worry', in R. G. Fox and B. J. King (eds), *Anthropology Beyond Culture.* Oxford: Berg, pp. 1–19.

Franklin, K. J. 1972. 'A Ritual Pandanus Language of New Guinea', *Oceania* 43: 66–76.

Freud, S. 1966. *The Standard Edition of the Complete Psychological Works of Sigmund Freud,* Vol. I. London: Hogarth.

Friedman, J. 2002. 'Modernity and Other Traditions', in B. M. Knauft (ed.), *Critically Modern: Alternatives, Alterities, Anthropologies.* Bloomington: Indiana University Press, pp. 287–313.

Friedman, J. and J. G. Carrier (eds). 1996. *Melanesian Modernities,* Lund Monographs in Social Anthropology No. 3. Lund: Lund University Press.

Gaines, A. D. 1992. 'Ethnopsychiatry: The Cultural Construction of Psychiatries', in A. D. Gaines (ed.), *Ethnopsychiatry: The Cultural Construction of Professional and Folk Psychiatries,* Albany, NY: State University of New York Press, pp. 3–49.

Ginzburg, C. 1983. *The Night Battles: Witchcraft and Agrarian Cults in the Sixteenth and Seventeenth Centuries.* London: Routledge and Kegan Paul.

Goddard, M. 1991. 'The Birdman of Kiripia: Posthumous Revenge in a Highland Papua New Guinea Society', in A. Pawley (ed.), *Man and a Half: Essays in Pacific Anthropology and Ethnobiology in Honour of Ralph Bulmer.* Auckland: JPS, pp. 218–22.

———. 1992. 'Bedlam in Paradise: A Critical History of Psychiatry in Papua New Guinea', *Journal of Pacific History* 27(1): 55–72.

———. 1994. 'A Suitable Case for Treatment? The Theory and Practice of Transcultural Psychiatry in Papua New Guinea', *Canberra Anthropology* 17(1): 30–52.

———. 1998. 'What Makes Hari Run? The Social Construction of Madness in a Highland Papua New Guinea Society', *Critique of Anthropology* 18(1): 61–81.

———. 2009. *Substantial Justice: An Anthropology of Village Courts in Papua New Guinea.* Oxford: Berghahn Books.

Golson, J. 1981. 'New Guinea Agricultural History: A Case Study', in D. Denoon and C. Snowden (eds), *A Time to Plant and a Time to Uproot: A History of Agriculture in Papua New Guinea.* Port Moresby: Institute of Papua New Guinea Studies, pp. 55–64.

———. 1997. 'The Tambul Spade', in H. Levine and A. Ploeg (eds), *Work in Progress: Essays in New Guinea Highlands Ethnography in Honour of Paula Brown Glick,* Frankfurt: Peter Lang, pp. 142–71.

Goody, J. 1987. *The Interface between the Written and the Oral.* Cambridge: Cambridge University Press.

Goody, J. and I. P. Watt. 1963. 'The Consequences of Literacy', *Comparative Studies in History and Society* 5: 304–45.

Gregory, C. A. 1982. *Gifts and Commodities.* London: Academic Press.

Guarnaccia, P. J. and L. H. Rogler. 1999. 'Research on Culture-Bound Syndromes: New Directions', *American Journal of Psychiatry* 156(9): 1322–27.

Hann, C. M. 2002. 'All *Kulturvölker* Now? Social Anthropological Reflections on the German-American Tradition', in R. G. Fox and B. J. King (eds), *Anthropology Beyond Culture.* Oxford, Berg, pp. 259–76.

Hasluck, P. 1976. *A Time For Building.* Melbourne: Melbourne University Press.

Head, J. 1990. 'Two Verbal Constructions in Kaugel', *Language and Linguistics in Melanesia* 21: 99–121.

———. 1993. 'Observations on Verb Suffixes in Umbu-Ungu', *Language and Linguistics in Melanesia* 24: 63–72.

Head, R. 1974. 'Gawigl', in K. A. McElhanon (ed.), *Legends From Papua New Guinea,* Ukaraumpa: Summer Institute of Linguistics, pp. 91–102.

Hide, R. 2003. *Pig Husbandry in New Guinea: A Literature Review and Bibliography.* Canberra: Australian Centre for International Agricultural Research.

Hirsch, E. 2001. 'When was Modernity in Melanesia?', *Social Anthropology* 9(2): 131–46.

Hobsbawm, E. J. (ed.). 1964. *Karl Marx: Precapitalist Economic Formations.* New York: International Publishers.

Hughes, C. C. 1998. 'The Glossary of "Culture-Bound Syndromes" in DSM-IV', *Transcultural Psychiatry* 35(3): 413–21.

Hughes, H. 2004. 'Can Papua New Guinea Come Back from the Brink?', *Issue Analysis* 49: 1–12.

Ingold, T. 2002. 'Introduction to Culture', in T. Ingold (ed.), *Companion Encyclopedia of Anthropology.* London: Routledge, pp. 329–49.

Jebens, H. 2005. *Pathways to Heaven: Contesting Mainline and Fundamentalist Christianity in Papua New Guinea.* London: Berghahn Books.

Johnson, F. Y. A. 1997. 'Ward Six Psychiatric Unit at the Port Moresby General Hospital: A Historical Review and Admission Statistics from 1980 to 1989', *PNG Medical Journal* 40(2): 79–88.

Jolly, M. 2001. 'Damning the Rivers of Milk? Fertility, Sexuality and Modernity in Melanesia and Amazonia', in T. A. Gregor and D. Tuzin (eds), *Gender in Amazonia and Melanesia: An Exploration of the Comparative Method.* Berkeley: University of California Press, pp. 175–207.

Karoma, J. F. 1978. 'The Origin and the Formation of the Korika Kengelka Tribe in the Tambul Sub-Province of W.H.P.', *Oral History* 6(1): 33–66.

Kettle, E. 1979. *That They Might Live.* Sydney: F. P. Leonard.

Kiev, A. 1972. *Transcultural Psychiatry.* New York: Free Press.

Kirmayer, L. J. 1998. 'The Fate of Culture in DSM-IV', *Transcultural Psychiatry* 35(3): 339–42.

———. 2006. 'Beyond the "New Cross-Cultural Psychiatry": Cultural Biology, Discursive Psychology and the Ironies of Globalization', *Transcultural Psychiatry* 43(1): 126–44.

Knauft, B. L. 1997. 'Gender Identity, Political Economy and Modernity in Melanesia and Amazonia', *Journal of the Royal Anthropological Institute* 3(2): 233–59.

———. 2002. *Exchanging the Past: A Rainforest World of Before and After.* Chicago: University of Chicago Press.

Koka, B. E., F. P. Deane and G. Lambert. 2004. 'Health Worker Confidence in Diagnosing and Treating Mental Health Problems in Papua New Guinea', *South Pacific Journal of Psychology* 15(1): 29–42.

Kraepelin, E. 1919. *Dementia Praecox.* Edinburgh: E. and S. Livingstone.

Kroeber, A. L. and C. Kluckhohn. 1952. *Culture: A Critical Review of Concepts and Definitions,* Cambridge, Mass.: Peabody Museum of American Archaeology and Ethnology, Harvard University.

Kuper, A. 1999. *Culture: The Anthropologist's Account.* Cambridge: Harvard University Press.

Laing, R. D. 1973. *The Politics of Experience/The Bird of Paradise.* London: Penguin.

Laing, R. D. and A. Esterton. 1970. *Sanity, Madness and the Family.* Middlesex: Penguin.

Langness, L. L. 1965. 'Hysterical Psychosis in the New Guinea Highlands: A Bena Bena Example', *Psychiatry* 28: 258–77.

Larrain, J. 1986. *A Reconstruction of Historical Materialism.* London: Allen and Unwin.

Leahy, M. 1936. 'The Central Highlands of New Guinea', *The Geographical Journal* 87(3): 229–60.

Lebra, W. P. (ed.). 1976. *Culture-Bound Syndromes, Ethnopsychiatry, and Alternative Therapies.* Honolulu: University of Hawai'i Press.

Leff, J. 1981. *Psychiatry Around the Globe: A Transcultural View.* New York: Marcel Dekker.

Lepi, P. and N. Bowers (eds). 1983. *Kaugel Stories: Temane and Kange,* Port Moresby: Institute of Papua New Guinea Studies, Oral History 11(4).

Lesser, A. 1961. 'Social Fields and the Evolution of Society', *Southwestern Journal of Anthropology* 17: 40–48.

Levack, B. P. 1987. *The Witch-Hunt in Early Modern Europe*. London: Longman.

Lévi-Strauss, C. 1968. *The Savage Mind*. London: Weidenfeld and Nicolson.

———. 1972. *Structural Anthropology*. Norwich: Penguin.

Lewis-Fernandez, R. 1998. 'A Cultural Critique of the DSM-IV Dissociative Disorders Section', *Transcultural Psychiatry* 35(3): 387–400.

LTP [Laws of the Territory of Papua]. 1949. *Laws of the Territory of Papua 1888–1945, Vol III*. Sydney: Halstead Press Pty Ltd.

MacDonald, M. 1979. 'Madness and Healing in Seventeenth-Century England'. PhD Thesis, Stanford University.

Macfarlane, A. 1970. *Witchcraft in Tudor and Stuart England*. London: Routledge and Kegan Paul.

Macgregor, D. F. 1967. 'Notes of a Meeting between Three South Pacific Psychiatrists', in *Mental Health in the South Pacific*, Noumea: South Pacific Commission, pp. 5–6.

Malinowski, B. 1966. *Argonauts of the Western Pacific*. New York: Dutton.

Malone, D. 2004. *The In-Between People: Language and Culture Maintenance and Mother-Tongue Education in the Highlands of Papua New Guinea*. Dallas: SIL International.

Mandrou, R. 1979. *Possession et Sorcellerie au XVIIe Siecle*. Paris: Fayard.

Marx, K. 1973. *Grundrisse*. Middlesex: Penguin.

———. 1974. *Economic and Philosophic Manuscripts of 1844*. Moscow: Progress Publishers.

Marx, K. and F. Engels. 1976. *The German Ideology*. Moscow: Progress Publishers.

May, R. J. 2001. *State and Society in Papua New Guinea: The First Twenty-Five Years*. Adelaide: Crawford House Publishing.

Mellett, D. J. 1982. *The Prerogative of Asylumdom: Social, Cultural and Administrative Aspects of the Institutional Treatment of the Insane in Nineteenth-Century Britain*. New York: Garland.

Mental Disorders Ordinance. 1960. Port Moresby: Govt Printer.

Merlan, F. and A. Rumsey. 1991. *Ku Waru: Language and Segmentary Politics in the Western Nebilyer Valley, Papua New Guinea*. Cambridge: Cambridge University Press.

Midelfort, H. C. E. 1980. 'Madness and Civilization in Early Modern Europe: A Reappraisal of Michel Foucault', in B. C. Malament (ed.), *After the Reformation: Essays in honour of J. E. Hexter*, Philadelphia: University of Pennsylvania Press, pp. 247–65.

Miyaji, N. T. 2002. 'Shifting Identities and Transcultural Psychiatry', *Transcultural Psychiatry* 39(2): 173–95.

Moi, W. 1976. 'Primary Prevention of Mental Disorder in Papua New Guinea'. Unpub. address to World Psychiatric Association, Tahiti. Mimeo, Dept of Health, Port Moresby, Papua New Guinea.

Monckton, A. W. 1922. *Last Days in New Guinea*. London: Bodley Head.

Neugebauer, R. 1981. *Mental Illness and Government Policy in Sixteenth and Seventeenth Century England*. Ann Arbor: University of Michigan Press.

Noricks, J. 1981. 'The Meaning of Niutao "Fakavalevale" (crazy) Behaviour: A Polynesian Theory of Mental Disorder', *Pacific Studies* 5(1):19–33.

Okpaku, S. O. (ed.). 1998. *Clinical Methods in Transcultural Psychiatry*. Washington, DC: American Psychiatry Press.

Orotoloa, P., T. Williams and G. de Moore. 2006. 'To be the First Psychiatrist in the Solomon Islands', *Australasian Psychiatry* 14(3): 243–45.

Otto, T. and A. Borsboom (eds). 1997. *Cultural Dynamics of Religious Change in Oceania.* Leiden: KITLV Press.

Pawley, A. 1992. 'Kalam Pandanus Language: An Old New Guinea Experiment in Language Engineering', in T. Dutton, M. Ross and D. Tryon (eds), *The Language Game: Papers in Memory of Donald C. Laycock,* Pacific Linguistics, C-110. Canberra: Australian National University, pp. 313–34.

Pedersen, P. B., N. Sartorius and A. J. Marsella (eds). 1984. *Mental Health Services: The Cross-Cultural Context.* Beverly Hills: Sage.

Pilgrim, D. 2007. 'The Survival of Psychiatric Diagnosis'. *Social Science and Medicine* 65: 536–47.

Reay, M. 1959. *The Kuma.* Melbourne: Melbourne University Press.

———. 1960. '"Mushroom Madness" in the New Guinea Highlands', *Oceania* 31(1): 137–39.

———. 1965. 'Mushrooms and Collective Hysteria', *Australian Territories,* 5(1): 18–28.

———. 1977. 'Ritual Madness Observed: A Discarded Pattern of Fate in Papua New Guinea', *The Journal of Pacific History* 12(1): 55–79.

Robbins, J. 2004. *Becoming Sinners: Christianity and Moral Torment in a Papua New Guinea Society.* Berkeley, CA: University of California Press.

Robin, R. 1979. *Psychopathology in Papua New Guinea.* Port Moresby: University of PNG.

Roheim, G. 1950. *Psychoanalysis and Anthropology.* New York: International Universities Press.

———. 1962. *Magic and Schizophrenia.* Bloomington, IN: Indiana University Press.

Rosen, G. 1968. *Madness in Society: Chapters in the Historical Sociology of Mental Illness.* London: Routledge and Kegan Paul.

RPNG. 1972. *Report to the General Assembly of the United Nations on the Administration of Papua New Guinea,* 1 July 1970–30 June 1971. Canberra: Govt Printer.

———. 1973. *Report to the General Assembly of the United Nations on the Administration of Papua New Guinea,* 1 July 1971–30 June 1972. Canberra: Govt Printer.

RTNG. 1958. *Report to the General Assembly of the United Nations on the Administration of the Territory of New Guinea,* 1 July 1956–30 June 1957. Canberra: Govt Printer.

———. 1959. *Report to the General Assembly of the United Nations on the Administration of the Territory of New Guinea,* 1 July 1957–30 June 1958. Canberra: Govt Printer.

———. 1960. *Report to the General Assembly of the United Nations on the Administration of the Territory of New Guinea,* 1 July 1958–30 June 1959. Canberra: Govt Printer.

Sahlins, M. 1999. 'Two or Three Things that I Know about Culture', *Journal of the Royal Anthropological Institute* 5(3): 399–421.

Samana, U. 1974. 'What goes on at Alkena "Yangpela Didiman Senta"', in J. P. Powell and M. Wilson (eds), *Education and Rural Development in the Highlands of Papua New Guinea.* Port Moresby: University of Papua New Guinea, pp. 224–41.

Sartre, J-P. 1976. *Critique of Dialectical Reason.* London: New Left Books.

Scheper-Hughes, N. and A. M. Lovell (eds). 1987. *Psychiatry Inside Out: Selected Works of Franco Basaglia.* New York: Columbia University Press.

Schmidt. A. 1971. *The Concept of Nature in Marx.* London: New Left Books.

Schneider, D. M. 1984. *A Critique of the Study of Kinship.* Ann Arbor: University of Michigan Press.

Scull, A. T. 1979. *Museums of Madness: The Social Organization of Insanity in Nineteenth-Century England.* New York: St Martin's Press.

———, (ed.). 1981. *Madhouses, Mad-Doctors and Madmen: The Social History of Psychiatry in the Victorian Era.* London: Athlone Press.

Seixas, P. (ed.). 2004. *Theorizing Historical Consciousness*. Toronto: University of Toronto Press.

Simons, R. C. and C. C. Hughes (eds). 1985. *The Culture-Bound Syndromes*. Dordrecht: D. Reidel Publishing Co.

Sinclair, A. J. 1957. *Field and Clinical Survey Report of the Mental Health of the Indigenes of the Territory of Papua and New Guinea*. Port Moresby: Govt Printer.

Solomon, P. and V. D. Patch. 1974. *Handbook of Psychiatry*. Los Altos: Lange Medical Publications.

Stafford-Clarke, D. 1964. *Psychiatry for Students*. London: George Allen and Unwin.

Stewart, P. J. and A. Strathern. 2001. *Humors and Substances: Ideas of the Body in New Guinea*. Westport: Bergin and Garvey.

Strathern, A. J. 1971. *The Rope of Moka: Big-Men and Ceremonial Exchange in Mount Hagen, New Guinea*. Cambridge: Cambridge University Press.

―――. 1975a. 'Why is Shame on the Skin?' *Ethnology* 14(4): 347–56.

―――. 1975b. 'Veiled Speech in Mount Hagen', in M. Bloch (ed.), *Political Language and Oratory in Traditional Society*, London: Academic Press, pp. 185–204.

―――. 1981. '"Noman": Representations of Identity in Mount Hagen'. In L. Holy and M. Stuchlik (eds), *The Structure of Folk Models*, London: Academic Press, pp. 281–303.

Strathern, A. J. and P. J. Stewart. 2000. *Arrow Talk: Transaction, Transition, and Contradiction in New Guinea Highlands History*. Kent, OH: Kent State University Press.

Strathern, M. 1972a. *Women in Between*. London: Seminar Press.

―――. 1972b. *Official and Unofficial Courts: Legal Assumptions and Expectations in a Highlands Community*. New Guinea Research Bulletin No 47, New Guinea Research Unit. Canberra: Australian National University.

―――. 1974. 'Managing Information: The Problems of a Dispute Settler (Mount Hagen)', in A. L. Epstein (ed.), *Contention and Dispute: Aspects of Law and Social Control in Melanesia*. Canberra: Australian National University Press, pp. 271–316.

Strauss, H. 1990. *The Mi-Culture of the Mount Hagen People, Papua New Guinea*, ed. G. Stürzenhofecker and A. Strathern, trans. B. Shields. Ethnology Monograph 13. Pittsburgh, PA: University of Pittsburgh.

Sullivan, N. (ed.). 2004. *Governance Challenges for PNG and the Pacific Islands*. Madang: Divine Word University Press.

Szasz, T. S. 1972. *The Myth of Mental Illness*. London: Paladin.

―――. 1973. *The Age of Madness: The History of Involuntary Mental Hospitalization Presented in Selected Texts*. New York: Anchor Press.

TPAR. 1955. *Territory of Papua Annual Report*, 1 July 1953–30 June 1954. Canberra: Govt Printer.

―――. 1959. *Territory of Papua Annual Report*, 1 July 1957–30 June 1958. Canberra: Govt Printer.

―――. 1960. *Territory of Papua Annual Report*, 1 July 1958–30 June 1959. Canberra: Govt Printer.

―――. 1962. *Territory of Papua Annual Report*, 1 July 1960–30 June 1961. Canberra: Govt Printer.

―――. 1963. *Territory of Papua Annual Report*, 1 July 1961–30 June 1962. Canberra: Govt Printer.

―――. 1965. *Territory of Papua Annual Report*, 1 July 1963–30 June 1964. Canberra: Govt Printer.

————. 1967. *Territory of Papua Annual Report*, 1 July 1965–30 June 1966. Canberra: Govt Printer.

————. 1968. *Territory of Papua Annual Report*, 1 July 1966–30 June 1967. Canberra: Govt Printer.

————. 1969. *Territory of Papua Annual Report*, 1 July 1967–30 June 1968. Canberra: Govt Printer.

Treu, R. and W. Adamson. 2006. 'Ethnomycological Notes from Papua New Guinea'. *McIlvainea* 16(2): 3–10.

Trouillot, M-R. 2002. 'Adieu, Culture: A New Duty Arises', in R. G. Fox and B. J. Fox (eds), *Anthropology Beyond Culture*. Oxford: Berg, pp. 37–60.

Tuke, S. 1964. *Description of the Retreat, an Institution Near York for Insane Persons of the Society of Friends, by Samuel Tuke, 1813; With an Introduction and Notes by Richard Hunter and Ida Macalpine*. London: Dawsons of Pall Mall.

Turkle, S. 1978. *Psychoanalytic Politics: Freud's French Revolution*. New York: Basic Books.

Turner, V. 1969. *The Ritual Process*. Chicago: Aldine Atherton.

Tuzin, D. 1997. *The Cassowary's Revenge: The Life and Death of Masculinity in a New Guinea Society*. Chicago: The University of Chicago Press.

United Nations. 1951. *United Nations Visiting Missions to Trust Territories in the Pacific 1950, Report on New Guinea*. New York.

————. 1953. *United Nations Visiting Missions to Trust Territories in the Pacific 1953, Report on New Guinea*. New York.

————. 1956. *United Nations Visiting Missions to Trust Territories in the Pacific 1956, Report on New Guinea*. New York.

————. 1959. *United Nations Visiting Mission to the Trust Territories of Nauru, New Guinea and the Pacific Islands, 1959, Report on New Guinea*. New York.

Van Gennep, A. 1960. *The Rites of Passage*. Chicago: University of Chicago Press.

Van Heekeren, D. 2004. 'Feeding Relationship: Uncovering Cosmology in Christian Women's Fellowship in Papua New Guinea', *Oceania* 75(2): 89–108.

Vicedom, G. and H. Tischner. 1943–48. *Die Mbowamb*. 3 vols. Hamburg: Friedrichsen, De Gruyter and Co.

Wagner, R. 1978. *Lethal Speech: Daribi Myth as Symbolic Obviation*. London: Cornell University Press.

Waldram, J. B. 2006. 'The View From the Hogan: Cultural Epidemiology and the Return to Ethnography', *Transcultural Psychiatry* 43(1): 72–85.

Wartofsky, M. W. 1977. *Feuerbach*. Cambridge: Cambridge University Press.

Weiner, J. 1996. *The Lost Drum: The Myth of Sexuality in Papua New Guinea and Beyond*. Madison, WI: University of Wisconsin Press.

Westermeyer, J. (ed.). 1976. *Anthropology and Mental Health*. The Hague: Mouton.

WHO [World Health Organization]. 1975. *Country Health Information Profile: Papua New Guinea*. Manila: WHO.

————. 1981. *The Work of WHO in the Western Pacific Region 1 July 1979–30 June 1981*. Manila: WHO.

Williams, F. E. 1976. *The Vailala Madness and Other Essays*. London: C. Hurst and Co.

Wiessner, P., and A. Tumu. 1998. *Historical Vines: Enga Networks of Exchange, Ritual, and Warfare in Papua New Guinea*. Bathurst: Crawford House Publishing.

Wittkower, E. D. and R. H. Prince. 1976. 'Transcultural Psychiatry — An Overview', *Papua New Guinea Medical Journal* 9(1): 6–13.

Wolf, E. R. 1982. *Europe and the People Without History.* Berkeley, CA: University of California Press.

Wolfers, E. P. 1975. *Race Relations and Colonial Rule in Papua New Guinea.* Sydney: Australia and New Zealand Book Co.

Wurm, S. A. 1961. 'New Guinea Languages'. *Current Anthropology* 2(2): 114–16.

Young, A. 2008. 'A Time to Change our Minds: Anthropology and Psychiatry in the 21st Century', *Culture, Medicine and Psychiatry* 32: 298–300.

Zilboorg, G. and G. W. Henry. 1941. *A History of Medical Psychology,* New York: W. W. Norton and Co.

INDEX